Faouzi Derbel, Nabil Derbel, Olfa Kanoun (Eds.)
**Communication, Signal Processing & Information Technology**

# Advances in Systems, Signals and Devices

—

Edited by
Olfa Kanoun, University of Chemnitz, Germany

## Volume 4

# Communication, Signal Processing & Information Technology

Edited by
Faouzi Derbel, Nabil Derbel, Olfa Kanoun

**DE GRUYTER**
OLDENBOURG

**Editors of this Volume**

Prof. Dr.-Ing. Faouzi Derbel
Leipzig University of Applied Sciences
Chair of Smart Diagnostic and Online Monitoring
Wächterstrasse 13
04107 Leipzig, Germany
faouzi.derbel@htwk-leipzig.de

Prof. Dr.-Ing. Olfa Kanoun
Technische Universität Chemnitz
Chair of Measurement and Sensor Technology
Reichenhainer Strasse 70
09126 Chemnitz
olfa.kanoun@etit.tu-chemnitz.de

Prof. Dr.-Eng. Nabil Derbel
University of Sfax
Sfax National Engineering School
Control & Energy Management Laboratory
1173 BP, 3038 SFAX, Tunisia
n.derbel@enis.rnu.tn

ISBN 978-3-11-044616-6
e-ISBN (PDF) 978-3-11-044839-9
e-ISBN (EPUB) 978-3-11-043618-1
Set-ISBN 978-3-11-044840-5
ISSN 2365-7493
e-ISSN 2365-7507

**Library of Congress Cataloging-in-Publication Data**
A CIP catalog record for this book has been applied for at the Library of Congress.

**Bibliographic information published by the Deutsche Nationalbibliothek**
The Deutsche Nationalbibliothek lists this publication in the Deutsche Nationalbibliografie;
detailed bibliographic data are available on the Internet at http://dnb.dnb.de.

© 2017 Walter de Gruyter GmbH, Berlin/Boston
Typesetting: Konvertus, Haarlem
Printing and binding: CPI books GmbH, Leck
♾ Printed on acid-free paper
Printed in Germany

www.degruyter.com

# Preface of the Volume Editor

The fourth volume of the Series "Advances in Systems, Signals and Devices" (**ASSD**), contains international scientific articles devoted to the field of communication, signal processing and information technology. The scope of the volume encompasses all aspects of research, development and applications of the science and technology in these fields. The topics include information technology, communications systems, digital signal processing, image processing, video processing, image and video compression, modulation and signal design, content-based video retrieval, wireless and optical communication, technologies for wireless communication systems, biometry and medical imaging, adaptive and smart antennas, data fusion and pattern recognition, coding compression, communication for e-mobility, microwave active and passive components and circuits, cognitive and software defined radio, vision systems and algorithms.

These fields are addressed by a separate volume of the series. All volumes are edited by a special editorial board made up by renowned scientist from all over the world.

Authors are encouraged to submit novel contributions which include results of research or experimental work discussing new developments in the field of communication, signal processing and information technology. The series can be also addressed for editing special issues for novel developments in specific fields. Guest editors are encouraged to make proposals to the editor in chief of the corresponding main field.

The aim of this international series is to promote the international scientific progress in the fields of systems, signals and devices. It provides at the same time an opportunity to be informed about interesting results that were reported during the international SSD conferences.

It is a big pleasure of ours to work together with the international editorial board consisting of renowned scientists in the field of communication, signal processing and information technology.

The Editors
Faouzi Derbel, Nabil Derbel and Olfa Kanoun

# Advances in Systems, Signals and Devices

## Series Editor:

Prof. Dr.-Ing. Olfa Kanoun
Technische Universität Chemnitz, Germany.
olfa.kanoun@etit.tu-chemnitz.de

## Editors in Chief:

### Systems, Automation & Control

Prof. Dr.-Eng. Nabil Derbel
ENIS, University of Sfax, Tunisia
n.derbel@enis.rnu.tn

### Power Systems & Smart Energies

Prof. Dr.-Ing. Faouzi Derbel
Leipzig Univ. of Applied Sciences, Germany
faouzi.derbel@htwk-leipzig.de

### Communication, Signal Processing & Information Technology

Prof. Dr.-Ing. Faouzi Derbel
Leipzig Univ. of Applied Sciences, Germany
faouzi.derbel@htwk-leipzig.de

### Sensors, Circuits & Instrumentation Systems

Prof. Dr.-Ing. Olfa Kanoun
Technische Universität Chemnitz, Germany
olfa.kanoun@etit.tu-chemnitz.de

## Communication, Signal Processing & Information Technology

## Sensors, Circuits & Instrumentation Systems

# Contents

# Contents

M. Rößler, J. Langer and U. Heinkel

# Finding an Optimal Set of Breakpoint Locations in a Control Flow Graph

**Abstract:** With the advance of high-level synthesis methodologies it has become possible to transform software tasks, typically running on a processor, to hardware tasks running on a FPGA device. Furthermore, dynamic reconfiguration techniques allow dynamic scheduling of hardware tasks on an FPGA area at runtime. Combining these techniques allows dynamic scheduling across the hardware-software boundary. However, to interrupt and resume a task, its context has to be identified and stored. We propose a method to find an optimal set of breakpoints in the control flow of a hardware task, such that the introduced resource overhead for context access is minimized and a maximum latency between interrupt request and the end of the context storing is guaranteed. This set of breakpoints allows the context to be restricted to the essential subset of data. Our method opens the door to flexible task scheduling not only on one reconfigurable device but also between different devices and even software instances of the same task.

**Keywords:** High level synthesis, Hardware software codesign, Preemption, Heterogeneous systems.

# 1 Introduction

This paper contributes to the vision of a system consisting of many different computing devices, including various types of processing units (generic multicore processors, digital signal processors or graphics processing units) running software tasks and reconfigurable FPGA devices running hardware tasks. Today, the advance of high-level synthesis (HLS) techniques blur the distinction between the two worlds as it becomes possible to create implementations of a single design specification for all types of devices mentioned before. Typically, the input specification is an algorithmic program written in common programming languages such as C or derived variants.

**M. Rößler, J. Langer and U. Heinkel:** Chair Circuit and System Design, Department of Electrical Engineering and Information Technology, Chemnitz University of Technology, Chemnitz, Germany, emails: marko.roessler@etit.tu-chemnitz.de, jan.langer@etit.tu-chemnitz.de, ulrich.heinkel@etit.tu-chemnitz.de.

De Gruyter Oldenbourg, ASSD – Advances in Systems, Signals and Devices, Volume 4, 2017, pp. 1–16.
DOI 10.1515/9783110448399-001

The big advantage of modern computing systems is concurrency. In order to gain efficiency the operating system can interrupt the execution of a task, save its context to the memory, restore it on another processor and resume its execution. This approach introduces flexibility to systems that can react to changing task priorities and environment requirements at runtime while consuming fewer resources than they virtually provide.

In order to implement such scheduling schemes not only on homogeneous systems of one or more equal processors but also in heterogeneous systems of different processors and reconfigurable hardware devices, it is necessary to interrupt hardware tasks during execution and save their context in an device-independent way while minimizing the required resource overhead. In this work we do not focus on possible implementations of such systems as SoC or HPC including their respective pros and cons. The focus is rather on how to find an optimal set of states in the central control flow of a hardware task that represent locations to interrupt the execution. We call those states *breakpoints* (the terms preemption point or switch point are also used in the literature).

When an interrupt occurs, the control flow progresses until it reaches a breakpoint. Then it stores the position in the control flow (the state itself) and all variables and structured memories that have been assigned a defined value (write access) and might be needed in the future (read access). This is called the context of a breakpoint and will later be used to initialize a software or hardware instance of the same task that is targeted on a different device. The set of *live* variables and structured memories are that part of the context that differs in size from breakpoint to breakpoint. In addition to minimizing the resource overhead of a set of breakpoints, we want to guarantee a maximum latency of the interrupt request, i.e. the time between the interrupt and the completion of storing the context does not exceed a certain number of cycles.

It is important to stress that interruption in the context of this work addresses breakpoints in the control flow that are common in the software and hardware implementation of a task. The traditional way of interrupting/restoring task execution on a processor at almost any position in the control flow is not applicable. Our definition of maximum latency assumes a continuous task execution for the software world until a breakpoint or the end is reached. In fact this assumption does hold for the addressed heterogeneous systems. With the contrast of multiple orders of magnitude between current reconfiguration latencies and the switching latency of a general purpose processor running a thin operating system layer it remains a reasonable approximation.

# 2 Related work

Current research focuses not only on the problem of converting classic C-style programs into corresponding hardware instances, but also develops new languages and methodologies to write specifications that target many devices ranging from multi-core processors over GPUs to hardware [1, 2] as an enabling technology for a single source design environment to heterogeneous systems.

Preempting hardware tasks has long been proposed in the context of FPGAs or application specific programmable processors (ASPP). With FPGAs the read back approach for saving and storing the context was deployed in different ways, for example in [3, 4]. These methods operate on bitstream level and involve no additional hardware or design costs as the built-in hardware structure of the circuit is employed. [5] adresses the drawback of using the entire bitstream by filtering the state information from the stream. This requires very detailed information on the circuitry and introduces large dependencies on the FPGA vendors. However, all bitstream based methods do not allow context switching between software and hardware tasks.

There has been effort to introduce specific structures to hold state information on higher levels. In [6] a register scan chain approach including a preemption controller is proposed. It is shown that preemptive flip flops introduce low overhead in hardware but a high latency while transferring context because of the chaining. The search for an optimal set of states in which a scheduled control flow can be preempted is considered in [7] on the example of ASPP. Valid states are greedily chosen and refined in a second step to minimize a cost function describing the hardware overhead. Nevertheless, it remains unclear whether this approach can handle general control flow including indefinite loops and branches.

The problem of interrupting hardware tasks at selected breakpoints, saving its context and resuming its operation either as a software task or as a hardware task at a different device has been studied in [8]. However, this approach requires the developer to explicitly define all breakpoint locations and determine the variables of the breakpoint's context in the C code using specific annotations. Optimizing the set of living variables by rescheduling HLS designs, has been addressed in [9]. They minimize state retention registers that are related to power optimizations.

# 3 Control flow graph

A high-level description of the design is given as an algorithmic program. This program is transformed into hardware by a HLS tool resulting in the generation of a scheduled control flow graph (CFG) representing the finite state machine of the hardware and a data flow graph (DFG) connecting operations and variables. With the operations of the DFG being scheduled to control states and allocated to hardware resources, we know exactly which variable is read or written in which control state. For the sake of simplicity structured data and buffered input/output ports are treated as a single variable. By means for any future read access the respective memory resource is involved in the state information. In the following, we introduce the terminology and some definitions.

The *control flow graph CFG* = $(S, E)$ consists of a set of *control states* $S = \{s_1, s_2, \ldots, s_n\}$ as its vertices and a set of transitions $E \subseteq S \times S$ between control states as its edges. The graph contains exactly one start state that has no predecessors and one end state that has no successors.

A *path* $p = (s_1, s_2, \ldots, s_k)$ is an ordered sequence of control states starting in state $s_1$ and ending in state $s_k$ with every state pair $(s_i, s_{i+1}) \in E$ being a valid transition in the CFG. For all states $s$ of the CFG, there exists at least one path from the start state to $s$ and from $s$ to the end state. The set of all valid paths in the CFG is denoted as $P$.

Furthermore, let $V = \{v_1, v_2, \ldots, v_m\}$ be the set of variables of the DFG and let $R \subseteq S \times V$ relate a control state to all the variables that are read in this state. Accordingly, let $W \subseteq S \times V$ be the relation for write accesses. If a variable is read and written in the same state $((s, v) \in R \cap W)$, it is first read and then written. Additionally, we assume that every read access to a variable has been preceded by a write access to this variable in a previous state. Finally, every variable is assigned a bit width that is used to calculate the effort that is needed to process and store this variable *bitwidth* : $V \rightarrow \mathbb{N}$.

## Example

**Tab. 1.** Example C code of a loop reading *MAX* subsequent values from *input_x*, calculating an output value and writing it to *output_y*.

```
uint32 x,y;
uint8 i = 0;

do {
x = read(input_x);
y = 150*x*x + 130*x+125;
write(output_y, y);
i++;
} while (i<MAX);
```

In Tab. 1 a small code fragment is given that is used as a running example throughout this paper. It consists of a loop that reads subsequent values from an input port *input_x*, performs the calculation $y = 150x^2 + 130x + 125$ and writes the result to the output port *output_y*.

The resulting CFG and DFG generated by the HLS are given in Fig. 1. The DFG on the right hand side is vertically aligned such that all operations are drawn on the same

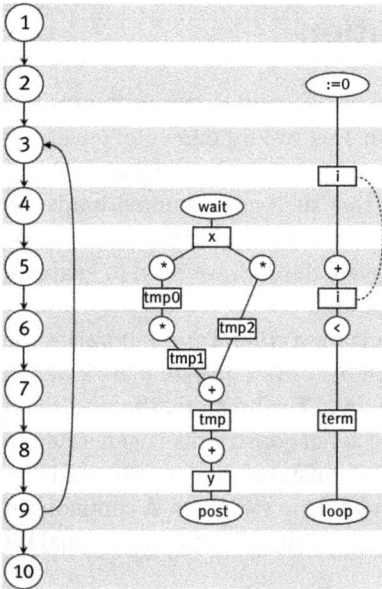

**Fig. 1.** Control flow graph (left) and data flow graph (both right graphs) of the running example.

line as the control state they are scheduled in. Of the two independent graphs the DFG consists of, the one in the center represents the calculation of $y$ from $x$, starts with reading the input (function *wait*) and finishes with writing the output (function *post*). The rightmost graph updates the loop counter and sets a variable *term* that controls the loop termination in state $s_9$. Furthermore, it can be seen that the synthesis tool introduces additional variables for temporary results ($tmp, tmp0, tmp1$ and $tmp2$). The initialization of variable $i$ occurs in state $s_2$.

We can derive the set of variables and the relations $R$ and $W$ of all read and write accesses:

$$V = \{i, term, tmp, tmp0, tmp1, tmp2, x, y\}$$
$$R = \{(s_5, x), (s_5, i), (s_6, tmp0), (s_6, i), (s_7, tmp1),$$
$$(s_7, tmp2), (s_8, tmp), (s_9, y), (s_9, term)\}$$
$$W = \{(s_2, i), (s_4, x), (s_5, tmp0), (s_5, tmp2), (s_5, i),$$
$$(s_6, tmp1), (s_6, term), (s_7, tmp), (s_8, y)\}.$$

Although in state $s_5$ variable $x$ is read by two operations, relation $R$ contains only one tuple $(s_5, x)$.

# 4 Problem formulation and solution

This paper proposes a method to find the optimal set of breakpoints in the control flow of a hardware description. In order to find such a set two fundamental questions must be answered. First, we need to define what exactly is meant by the term optimal. A good optimization criteria considers several factors such as additional hardware resources and expected latency between the request for a breakpoint and the time the context of the breakpoint has been successfully saved. Second, we need to examine which constraints must be met by the set of breakpoints.

In the following, we introduce a concrete optimization criteria that minimizes the hardware resource overhead by minimizing the number of variables that are saved at any breakpoint and the overall number of breakpoints. Furthermore, the maximum latency is set to a finite value and a corresponding set of constraints is generated to guarantee it. Finally, the optimization criteria and the constraints are transformed into an integer linear programming (ILP) problem with boolean variables. A common 0-1 ILP solver is used to find an optimal solution of the problem and hence an optimal set of breakpoints.

## 4.1 Context calculation

In order to stop and resume a hardware component at a specific breakpoint state, its context information must be extracted. The context at breakpoint $s$ consists of the values of all variables that have been assigned a value and are potentially read in the future

$$context : S \to 2^V.$$

This problem is similar to the well-understood liveness analysis of variables being employed by software compilers [10]. In contrast to compilers our data flow analysis operates on control states instead of code blocks.

A variable $v$ is part of the context of control state $s$, if there exists a path from state $s$ to an arbitrary state $s'$ where $v$ is read. There must be no write states on this path. If state $s$ itself is a write state, $v$ is part of the context, because in case of a breakpoint at this state the context is stored after the variable was written. On the other hand, the variable is not in the context of the last read state in the control flow, because after all the state's operations are completed, there won't be further need to store the variable. The context is computed with the following equation

$$context(s) = \left\{ v \in V \mid \exists (s p s') \in P \ / \ (s', v) \in R \ \wedge \ \bigwedge_p^{s_i} (s_i, v) \notin W \right\}$$

where $(sps')$ is a path starting at $s$ and ending in $s'$ with $p$ being (the possibly empty) subpath of all intermediate states.

For future reference, we define the size of a context as the number of bits in all variables of the context

$$csize(s) = \sum_{context(s)}^{v} bitwidth(v).$$

## 4.2 Optimization criteria

The goal of the optimization is the reduction of the hardware overhead that results from inserting the breakpoints and the latency between breakpoint request and the corresponding reaction of the hardware. First, we examine the different components of the hardware overhead. In this work, we do not propose a concrete implementation of the breakpoint hardware, but rather concentrate on the general influence of different factors.

The hardware resource overhead for a given set of breakpoints $BP \subseteq S$ consists of three factors:

- The number of breakpoints, because there is a static overhead for each breakpoint that is inserted in the control flow

$$cost_1(BP) = |BP|. \tag{1}$$

- The total number of bits in all variables that are used in any breakpoint, because each affected register must be changed to be able to store and restore its content in case of a breakpoint

$$cost_2(BP) = \sum_{V_{BP}}^{v} bitwidth(v) \tag{2}$$

where

$$V_{BP} = \bigcup_{BP}^{s} context(s).$$

- Furthermore, for every breakpoint there will be an overhead that depends on the number of bits to be stored in this breakpoint

$$cost_3(BP) = \sum_{BP}^{s} csize(s). \tag{3}$$

In contrast to the hardware overhead, the latency between away the breakpoint request and the time step the context is completely saved to some internal or external memory location consists of two parts:
- First, the time difference between the current time step and the next breakpoint.
- Second, as soon as the breakpoint is reached the context is stored. The duration of this process corresponds to the number of bits in the context of this state.

The exact value of the latency (or more correctly the probability distribution of the latency) depends on factors such as the probability of a breakpoint request in each control state and the sequence of input values that control the actual sequence of control states. Therefore, we concentrate on restricting our set of breakpoints to those sets that guarantee an upper bound for the latency. The next subsection describes how these constraints are obtained.

With respect to the overall optimization problem our proposed method minimizes the hardware overhead of a set of breakpoints while guaranteeing a maximum latency. The final cost function weighs and adds the three individual cost functions for the hardware overhead:

$$cost(BP) = w_1 \cdot cost_1(BP) + w_2 \cdot cost_2(BP) +$$
$$w_3 \cdot cost_3(BP). \tag{4}$$

The exact values of the weights $w_1$, $w_2$ and $w_3$ depend on the actual hardware implementation, i.e. how much hardware resource overhead must be introduced into the design in order to establish the breakpoint itself ($w_1$), to use a variable in any breakpoint ($w_2$) and to store the values of the variables of the breakpoint ($w_3$).

## 4.3 Constraining the maximum latency

The maximum latency requirement can be restated such that from any control state there exists no path that
- is longer than the maximum latency $L$ without reaching a breakpoint or
- the length of the path until the breakpoint plus the latency that is caused by the context storing exceeds $L$.

This implies that all indefinite loops contain at least one breakpoint, because otherwise the control flow can execute the loop as long as it takes to violate the maximum latency. In case of loops with a fixed maximum number of executions they need to be analyzed in fully unrolled manner distinguish whether a breakpoint inside the loop is required.

Our proposed algorithm for generating constraints that guarantee a maximum latency as shown in Tab. 2 first iterates over all control states. For each state $s$ we extract all paths with a maximum length $L$. Traversing is stopped if a state occurs twice in the path, because that indicates a loop and every loop must include at least one breakpoint. The set $Q$ contains all partly traversed paths. Newly discovered paths are added to the set whereas paths that reached the limit are moved into the set *constraints*. This set is also the result of the algorithm.

**Tab. 2.** Scheme to extract all constraints by enumerating all paths of length $\leq L$ starting from any control state $s$.

```
constraints ⟵ ∅
for all s ∈ S
Q ⟵ {(s)}
   while Q ≠ ∅
   p ⟵ choose from Q
   Q ⟵ Q \ {p}
     if |p| = L
        constraints ⟵ constraints ∪ {purge(p)}
     else
        x ⟵ last state in p
        succ ⟵ {y ∈ S | (x, y) ∈ E}
     if (succ ∩ p) ≠ ∅
        constraints ⟵ constraints ∪ {purge(p)}
     else
     for all y ∈ S | (x, y) ∈ E
           Q ⟵ Q ∪ {p + (y)}
     end for
     end if
     end if
   end while
end for
```

Additionally, the algorithm needs to consider the latency that is caused by storing the context of the breakpoint state. This latency is calculated for every control state separately as the sum of all bits in the context divided the bit transmission rate $R_b$ in bits per time step

$$latency(s) = \left\lceil \frac{1}{R_b} \cdot csize(s) \right\rceil .$$

Now we can define the helper function *purge* that has already been introduced in Alg. 2 as

$$purge(p) = \{s_i \in p \mid i + latency(s) \leq L\}$$

where $i$ represents the position of state $s$ in path $p$. The first state in $p$ is denoted by $s_0$.

Each set of states in *constraints* (due to the purging the sets are no longer valid paths) requires that at least one of the control states in the set must be a breakpoint.

**Example**

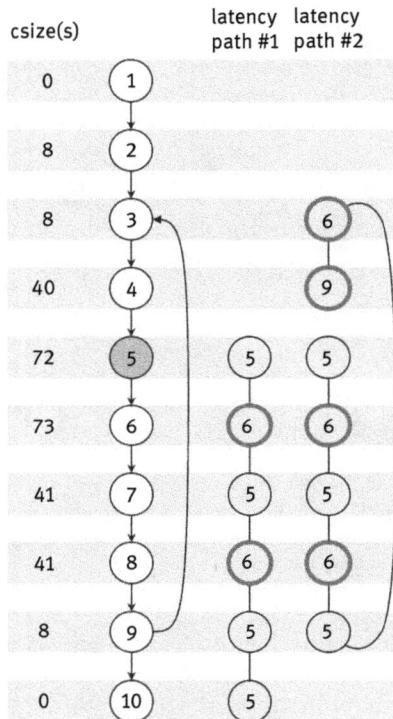

**Fig. 2.** CFG showing two possible paths starting from control state $s_5$.

The algorithm is best understood by the illustration in Fig. 2. It shows the CFG with state $s_5$ marked to indicate that this state is currently examined. We want to find out which constraints must be met in order to guarantee the maximum latency requirement in case a breakpoint request arrives at $s_5$. Starting from it, there are two possible computation paths of the CFG. The first one aborts the loop and finishes in state $s_{10}$. The second one executes the loop once more and stops at state $s_5$ because after that the initial state is reached again. The first column displays the number of bits in the context of the corresponding state. The bit width of variable $i$ is 8, *term* is a single bit variable and all other variables have 32 bit. The states of the two paths are labeled with the total latency for the case this state is a breakpoint. The bit transmission rate is $R_b$ = 16 and we want to guarantee a maximum latency of 5 time steps.

All states whose total latency exceeds the requirement of 5 time steps are marked with a bold red border. The resulting constraints are

$$constraints = \{\ldots, \{s_5, s_7, s_9, s_{10}\}, \{s_5, s_7, s_9\}, \ldots\}.$$

We can remove the first constraint because it is covered by the second one. The corresponding boolean constraint for state $s_5$ is

$$(s_5 \in BP) \lor (s_7 \in BP) \lor (s_9 \in BP)$$

## 4.4 Solution

In order to find a solution to our optimization problem, we transform it into a pseudo-boolean (0-1) ILP problem. For every control state $s \in S$ we define a boolean ILP variable $x_s$. A value of 0 for this boolean variable means that the corresponding control state is no breakpoint, whereas a value of 1 means it is a breakpoint

$$x_s = \begin{cases} 1 & s \in BP \\ 0 & otherwise \end{cases}.$$

Accordingly, for every data flow variable $v \in V$ we also define a boolean ILP variable $x_v$, such that the boolean value of it indicates whether $v$ appears in the context of any breakpoint

$$x_v = \begin{cases} 1 & \exists s \in BP : v \in context(s) \\ 0 & otherwise \end{cases}.$$

All boolean variables are collected in the vector $\bar{x} = (x_{s_1}, \ldots, x_{s_n}, x_{v_1}, \ldots, x_{v_m})$. An ILP problem requires an objective function in the form of $\bar{c}^T \bar{x}$ whose result is to be minimized (or maximized). The elements of weight vector $\bar{c} = (c_{s_1}, \ldots, c_{s_n}, c_{v_1}, \ldots, c_{v_m})$ are the constants that are directly derived from the corresponding cost functions. From equations 1, 3 and 4 we can obtain the weights for all control states

$$c_s = w_1 + w_3 \cdot csize(s)$$

and all data flow variables

$$c_v = w_2 \cdot bitwidth(v).$$

So far, we did not introduce any connection between data flow variables $V$ and control states $S$, e.g. the solver can set all $x_s$ variables to 1 and all $x_v$ variables to 0 which is not a valid solution. Hence, a control state needs to imply all variables that belong to its context. This logical implication is transformed into regular ILP constraints

$$\bigwedge_{S}^{s} (x_s \rightarrow \bigwedge_{context(s)}^{v} x_v)$$

$$= \bigwedge_{V}^{v} (x_v \vee \bigwedge_{S_v}^{s} \neg x_s)$$

$$= \bigwedge_{V}^{v} (|S_v| \cdot x_v - \sum_{S_v}^{s} x_s \geq 0)$$

with $S_v = \{s \in S \mid v \in context(s)\}$.

The boolean constraints of the latency requirement are similarly converted to ILP constraints and all constraints that are covered by other constraints are removed

$$\bigwedge_{constraints\ c}^{c} \bigvee^{s} x_s$$

$$\cong \bigwedge_{constraints\ c}^{c} (\sum^{s} x_s \geq 1).$$

After choosing adequate values for $w_1, w_2$ and $w_3$, all constraints and the objective function are written and an ILP solver (not necessarily a pseudo-boolean ILP solver) can start to find an optimal solution.

## Example

After choosing arbitrary weights of $w_1 = 10$, $w_2 = 3$ and $w_3 = 1$, a bit rate $R_b = 16$ and a maximum latency of $L = 5$ we find an optimal set of breakpoints $BP = \{s_2, s_5, s_{10}, s_{11}\}$. It can be noted that the last control state is always selected as a breakpoint.

# 5 Results

The optimal set of breakpoints has been generated for four different designs. The input description language and the corresponding HLS tool used is the open source Streams-C implementation [11]. It has been chosen for historical reasons of the research group and does not provide leading edge optimization as recent compilers (ROCCC, LegUp or Gaut). Nevertheless, as this work was built on the general output of HLS, the basic set of transformations provided by Streams-C is sufficient for an analysis. The four designs are our running example used throughout this paper (*while_calc*), another simple loop example (*cascaded_for_if*), a Bayes filter implementation (*particle*) and an MJPEG en-/decoder (*MJPEG_coder*), which consists of six concurrent tasks.

**Tab. 3.** Results for four designs, run with parameters $w_1 = 10$, $w_2 = 3$ and $w_3 = 1$, a bit rate $R_b = 16$ and maximum latency $L$.

| Design | LOC | L | # BPs | # Vars | # Bits | Solver Runtime |
|---|---|---|---|---|---|---|
| *while_calc* | 9 | 5 | 4/10 (40%) | 2/8 (25%) | 40/201 (20%) | < 1s |
| *cascaded_for_if* | 16 | 5 | 3/11 (27%) | 4/10 (40%) | 66/165 (40%) | < 1s |
| *particle* | 68 | 40 | 11/130 (8%) | 10/120 (8%) | 243/2846 (8%) | 19s |
| *MJPEG_coder* | | | | | | |
| – *ColTrans* | 98 | 100 | 5/41 (12%) | 5/49 (10%) | 74/1008 (7%) | 3s |
| – *InvColTrans* | 100 | 100 | 4/50 (8%) | 3/47 (6%) | 18/938 (2%) | 2s |
| – *CosTrans* | 315 | 100 | 10/432 (2%) | 9/335 (3%) | 58/6869 (1%) | 104s |
| – *InvCosTrans* | 211 | 100 | 11/387 (3%) | 11/75 (15%) | 98/1040 (9%) | 50s |
| – *Huffman* | 515 | 100 | 17/648 (3%) | 11/60 (18%) | 97/669 (14%) | 57s |
| – *InvHuffman* | 533 | 100 | 10/567 (2%) | 0/19 (0%) | 0/332 (0%) | 3s |

In Tab. 3 we show the characteristic parameters for these problems and the results of our algorithm with the fixed maximum latency in column three. The second column contains the lines of C code of the pure algorithm body that is synthesized. Columns 4–6 show the number of breakpoints, variables used in any breakpoint and the corresponding number of bits. Additionally, the total number of control states, variables and bits is shown. Finally, the last column contains the total runtime of the ILP solver in seconds. The runtime of the problem generation does not exceed one second on any of the processed designs.

The pseudo-boolean ILP solver we used is the function `bintprog` from MATLAB's optimization toolbox. The experiments have been run on a 2.6GHz Dual-Core processor with 16GB memory.

In order to evaluate our method we examined the *particle* design and generated the optimal set of breakpoints for different values of $L$. In Fig. 3 the corresponding results are shown. For latencies smaller than 17 no solution could be found. Furthermore, for latencies greater than or equal to 32 no improvement in the number of breakpoints or estimated hardware overhead could be detected. It can be seen that the solver runtime varies greatly between ten seconds and about 17 minutes. The experiments show that the runtime of the solver tends to peak at values for $L$ that are slightly higher than the smallest $L$ for which a solution can be found. The runtime of the problem generation, and therefore our proposed algorithms, initially increases with $L$ but does not exceed 60ms even for very high $L$.

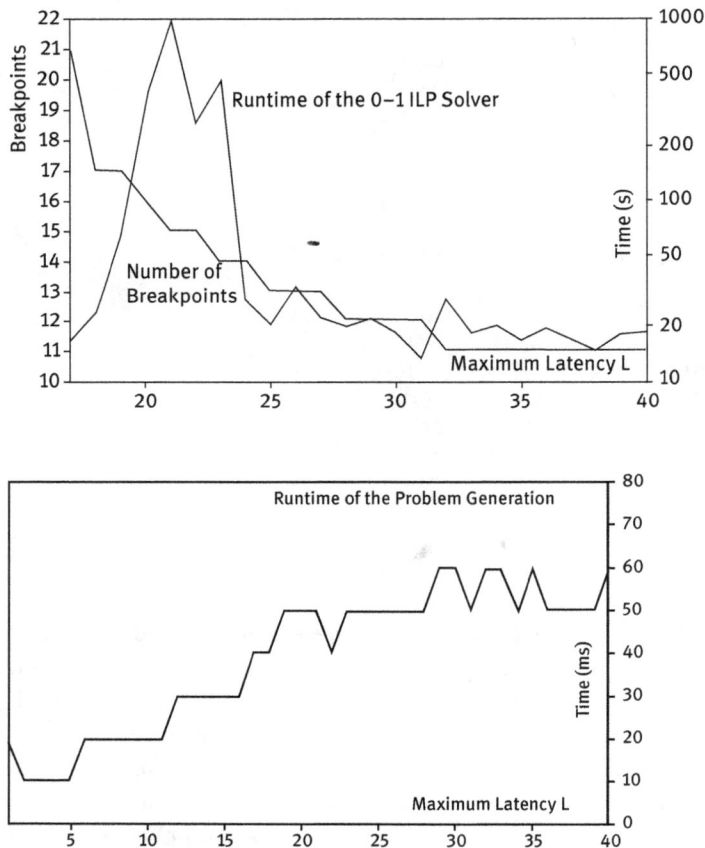

**Fig. 3.** Number of breakpoints and solver runtime for a range of values of the maximum latency $L$.

# 6 Conclusion

In computing systems of the next generation it will be desirable to switch processing tasks between software and hardware instances. For this it is necessary to interrupt a hardware task at runtime and save its context in order to resume its execution later on a different device. In this paper, we proposed an algorithm to find the optimal set of interrupt locations (breakpoints) which minimizes the required hardware overhead and holds a maximum latency between the interrupt request and completion of the context storing.

Currently, our algorithm does not consider additional information about the program structure such as loops of predefined length or branches of the control flow that are mutually exclusive. Including such information into the constraint calculation will help finding better breakpoint sets while still guaranteeing the maximum latency. Further improvement could be achieved by rescheduling the control and data flow by optimizing the set of living variables at proposed breakpoint states.

**Acknowledgment:** The work presented in this paper was done within the project InnoProfile GPSV and InnoProfile-Transfer GPZV which are funded by the german Federal Ministry of Education and Research (BMBF) within the program *Unternehmen Regionen* under contract numbers 03IP505 and 03IPT505X.

# Bibliography

[1]   J. Auerbach, D. F. Bacon, P. Cheng, and R. Rabbah, "Lime: a Java-Compatible and Synthesizable Language for Heterogeneous Architectures," in *Object Oriented Programming Systems Languages and Applications (OOPSLA)*. ACM, 2010, pp. 89–108. [Online]. Available: http://portal.acm.org/citation.cfm?doid=1869459.1869469

[2]   S. Singh, "Computing without Processors," *Communications of the ACM (CACM)*, vol. 54, no. 8, pp. 46–54, Aug. 2011. [Online]. Available: http://portal.acm.org/citation.cfm?doid=1978542.1978558

[3]   S. Trimberger, D. Carberry, A. Johnson, and J. Wong, "A Time-Multiplexed FPGA," in *Field-Programmable Custom Computing Machines (FCCM)*. IEEE, 1997, pp. 22–28. [Online]. Available: http://ieeexplore.ieee.org/lpdocs/epic03/wrapper.htm?arnumber=624601

[4]   H. Kalte and M. Porrmann, "Context Saving and Restoring for Multitasking in Reconfigurable Systems," in *Field Programmable Logic and Applications (FPL)*. IEEE, 2005, pp. 223–228. [Online]. Available: http://ieeexplore.ieee.org/lpdocs/epic03/wrapper.htm?arnumber=1515726

[5]   H. Simmler, L. Levinson, and R. Männer, "Multitasking on FPGA Coprocessors," in *Field Programmable Logic and Applications (FPL)*. IEEE, 2000, pp. 121–130.

[6]   S. Jovanovic, C. Tanougast, and S. Weber, "A Hardware Preemptive Multitasking Mechanism Based on Scan-path Register Structure for FPGA-based Reconfigurable Systems," in *NASA/ESA Conference on Adaptive Hardware and Systems (AHS)*. IEEE, Aug. 2007, pp. 358–364. [Online]. Available: http://ieeexplore.ieee.org/lpdocs/epic03/wrapper.htm?arnumber=4291942

[7]  K. Kim, R. Karri, and M. Potkonjak, "Micro-Preemption Synthesis: An Enabling Mechanism for Multi-Task VLSI Systems," in *Int. Conference on Computer Aided Design (ICCAD)*. IEEE, 1997, pp. 33–38.

[8]  M. Rößler and U. Heinkel, "Preemptive HW/SW-Threading by Combining ESL Methodology and Coarse Grained Reconfiguration," in *Reconfigurable Communication-centric Systems-on-Chip (ReCoSoC)*, 2008.

[9]  E. Choi, C. Shin, T. Kim, and Y. Shin, "Power-gating-aware high-level synthesis," in *Low Power Electronics and Design (ISLPED), 2008 ACM/IEEE International Symposium on*, aug. 2008, pp. 39–44.

[10]  B. Morgan, *Building an Optimizing Compiler*. Digital Press, 1997.

[11]  M. B. Gokhale, J. M. Stone, J. Arnold, and M. Kalinowski, "Stream-Oriented FPGA Computing in the Streams-C High Level Language," in *Field-Programmable Custom Computing Machines (FCCM)*.Washington, DC, USA: IEEE, 2000, p. 49.

# Biographies

**Marko Rößler** has completed his diploma in 2005 in Information and Communication Technology at Chemnitz University of Technology. He is pursuing his PhD degree from Chemnitz University of Technology since 2009. His areas of interest are hardware software codesign and high level synthesis of heterogenous systems.

**Jan Langer** received his Ph.D. in 2011 from Chemnitz University of Technology in Germany investigating the high-level synthesis of temporal properties. He is currently with the Xilinx Research Labs in Dublin, Ireland, where his research focus is on the software-programmability of wireless applications in the context of heterogeneous computing platforms.

**Ulrich Heinkel** is head of the chair Circuit and System Design at Chemnitz University of Technology. He received his diploma and PhD in electrical engineering from Friedrich-Alexander-University of Erlangen-Nuremberg. Since 1999 he worked as research engineer at Lucent Technologies Nuremberg. His research focusses on formal methods, specification and verification of digital systems.

L. Zimmermann, A. Goetz, G. Fischer and R. Weigel

# Performance Analysis of Time Difference of Arrival and Angle of Arrival Estimation Methods for GSM Mobile Phone Localization

**Abstract:** The localization methods time difference of arrival (TDOA) and angle of arrival (AOA) are investigated in their capability of high-resolution detection of GSM signals. Because GSM is of narrowband nature, a bandwidth expansion technique that significantly increases the resolution of TDOA is introduced. The algorithms maximum likelihood estimator (for TDOA), root multiple signal classification and estimation of signal parameters via rotational invariance techniques (for AOA) are proposed as a reference for each method. Simulations conducted with noisy signals and multipath propagation demonstrate the potential of TDOA for GSM localization, especially for search and rescue applications. AOA on the other hand is very sensitive to multipath, and is therefore only recommended for open space scenarios.

**Keywords:** Time difference of arrival estimation, Direction of arrival estimation, GSM, Maximum likelihood estimation, Multiple signal classification, Estimation of signal parameters via rotational invariance techniques.

## 1 Introduction

The research in high-resolution GSM positioning is part of a new localization system called "I-LOV" for search and rescue of buried people [1]. It is expected that at the time a building collapses, up to 80 % of the buried people carry along their mobile phone. Therefore, the I-LOV system intends to accelerate the search process by providing the rescue forces with possible positions of buried people based on network-based mobile phone localization. The decision to deploy GSM is due to the fact that (unlike 3G or 4G technologies) almost every mobile phone supports the standard. Apart from localizing an emergency call, GSM positioning is of interest in police and military applications, or may be utilized for location based services in smartphones.

To date, a variety of network-based GSM localization techniques are utilized. The most basic one is the method of cell identification (Cell-ID), which determines

**L. Zimmermann, A. Goetz, G. Fischer and R. Weigel:** Institute for Electronics Engineering, Friedrich-Alexander University Erlangen-Nuremberg, Erlangen, Germany, emails: lars.zimmermann@fau.de, christian.spratler@web.de, robert.weigel@fau.de, georg.fischer@fau.de, AGGoetz@web.de

De Gruyter Oldenbourg, ASSD – Advances in Systems, Signals and Devices, Volume 4, 2017, pp. 17–34.
DOI 10.1515/9783110448399-002

the mobile station's position by indicating the nearest base station. With the help of the timing advance value and by narrowing results down to the sector of the directional antenna, a positioning resolution of roughly 550 m can be achieved [2]. Other approaches are based on received signal strength measurements or multipath fingerprinting. These techniques, however, require complete knowledge of the radio channel, which is not feasible for search and rescue applications [3].

Another method called Time of Arrival (TOA), measures the time a signal travels from the mobile station to three or more neighboring base stations. The requirement here is the knowledge of the time the transmission begins as well as a synchronized time basis with the base stations [2]. Unfortunately, the former requirement can not be fulfilled as access to the mobile phone is generally not available.

In this paper, the high-resolution principles time difference of arrival (TDOA) and angle of arrival (AOA) are proposed (depicted in Fig. 1).

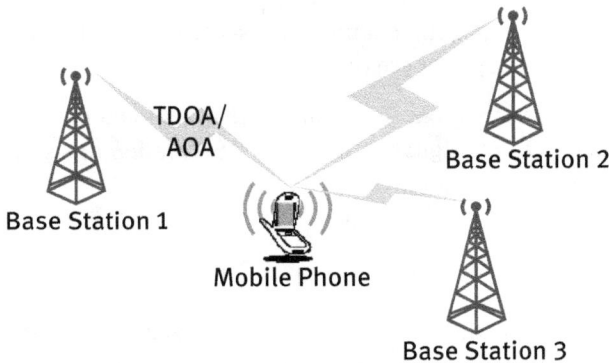

**Fig. 1.** Setup of TDOA and AOA estimation of a GSM mobile phone signal.

Both methods are compatible with the GSM standard, and do not require any knowledge on the radio channel or the transmission time of the mobile phone. Nevertheless, it is advisable to have full control over the base stations and AOA receivers require additional array antenna equipment. It is also possible to use operator-independent receiver stations [4].

Because results are highly dependent on the signal's bandwidth, our institute developed a bandwidth expansion technique for GSM that allows TDOA to reach a higher resolution. This makes it possible to compare the performance of TDOA in wideband with AOA in narrowband. The estimation algorithms presented here are the maximum likelihood estimator (MLE) for TDOA, root multiple signal classification (Root-MUSIC) and estimation of signal parameters via rotational invariance techniques (ESPRIT) for AOA. Although GSM has its own characteristics in signal composition and wave propagation, the results presented here may be transferable to other communication systems.

# 2 Signal and channel models

## 2.1 Global system for mobile communication

Despite the vast development in 3G and now 4G communication, GSM remains the world's most widespread mobile communication standard. This makes it a popular source for radio-location. Its specification is continuously updated by the 3rd Generation Partnership Project (3GPP) [5]. Europe and most other parts of the world operate in the frequency bands GSM 900 and 1800. Due to the cellular nature of the mobile systems, geographic areas hold only a subset of the available frequency bandwidth. Therefore, GSM combines two multiplexing techniques, frequency division multiple access (FDMA) and time division multiple access (TDMA).

The research in this paper focuses on the E-GSM 900 uplink band from 880–915 MHz that is spanned in 124 channels, each 200 kHz wide. The time axis is further divided in 8 (or 16 half-rate) time slots. One time slot covers 156.25 bits, which is equivalent to a duration of about 577 $\mu$s. There are five different data formats that can be used in a time slot. The normal burst, the most frequently used burst structure, is chosen as signal source as it typically transfers digitized voice data. Voice generates stochastically distributed signals, which is favorable for the proposed estimation techniques because the signal will only correlate with itself. In addition, the following channel coding, interleaving, encryption and modulation of the voice data further randomizes the signal. However, some bits are used for signaling purposes and add a deterministic part to the signal. GSM uses a continuous-phase frequency-shift keying modulation named Gaussian minimum shift keying (GMSK).

The simulation of the GSM normal burst in this paper is compliant to the GSM specifications. It includes a pseudo-random bit generator characterizing speech, a normal burst generator, channel coding, interleaving and a GMSK modulator [6]. As encryption does not change the random nature of the bits, it is not required here. A power ramping function on the other hand may affect the estimator's accuracy. The effects of power ramping is not studied here because it is expected to be negligible if the device is working accordingly to the standard, where tolerances for 147 of 148 bits are restricted to a ±1 dB power level. The multiplexing of the complex baseband signals is discussed in the following.

## 2.2 Dependency of resolution on signal bandwidth

The actual resolution of the proposed estimators is highly dependent on the signal's bandwidth. In general, the 200 kHz wide GSM signal can be declared narrowband. This, for instance, is beneficial for angle of arrival subspace methods, which are defined for narrowband signals and might fail in case of wideband signals [7].

This is different for the temporal resolution of the time difference of arrival method, which is proportional to the bandwidth of a signal. A GSM normal burst with a bandwidth of 200 kHz results in a coarse resolution of approximately 1.5 km (c/BW) [4]. In case of narrowband signals, estimators based on the correlation of signals have difficulties resolving the time delay between two consecutive signals. The result is characterized by a broadened correlation peak, which can only be resolved by a very high signal-to-noise-ratio (SNR) or exceedingly long observation intervals. Both criteria are usually bounded by the measurement equipment or the signal itself. The effect may become even more severe if multipath signals superimpose with the line-of-sight signal.

A technique within the GSM standard called frequency hopping is used to reduce frequency selective interferences caused by fading or interferences from adjacent channels. The base station triggers frequency hopping by switching the carrier frequency either periodic or pseudo-random after every TDMA frame. Figure 2 illustrates the time-frequency allocation with active frequency hopping. Here, the service provider's cell is assigned to three different frequency bands. The traffic channel hops after each frame to a predefined frequency channel.

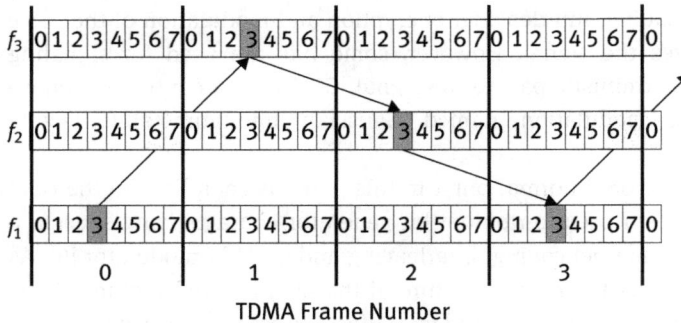

**Fig. 2.** Example of the time-frequency allocation with frequency hopping.

Research at our institute has demonstrated that by using frequency hopping, it is possible to virtually increase the frequency bandwidth on receiver side. This is realized by combining multiple bursts sent on varying carrier frequencies to one single signal [8]. For example in Fig. 2, a higher bandwidth signal is constructed by merging the first three TDMA frames. Then, the signal gets spanned over a bandwidth of 600 instead of 200 kHz (three times 200 kHz).

The bandwidth expansion allows a high temporal resolution for TDOA estimation. It can be applied if control over the base station is established, or at least the hopping sequence is known. Frequency hopping is simulated by a frequency mixer, which creates analytical signals on varying carrier frequencies of the E-GSM 900 band. In order to easily reconstruct the signal in the receiver, a fixed frequency spacing between following bursts is selected. The frequency mixer is followed by a multiplexer. In the simulations of this paper, the multiframe is simplified to contain only traffic channels.

## 2.3 Channel models

A line-of-sight (LOS) channel model is developed to analyze the effects of channel-induced noise on the performance of the proposed estimators, whereas the Wireless World Initiative New Radio (WINNER) II model is used to generate various multipath scenarios. Both models are geometry-based generic models that handle analytical as well as complex baseband signals. In order to create statistically independent events, each GSM burst is modulated with an initial random phase before channel modeling.

The LOS channel model is characterized by a line-of-sight signal with additive white Gaussian noise (AWGN). Modeled noise effects include the propagation path as well as the receiver system itself. The channel model draws the time and angle of arrival of a signal from the geometry provided by the user. The noise power of the 200 kHz signal is determined stochastically from the SNR. At present, a dipole antenna and a uniform linear array (ULA) antenna are supported.

In mathematical terms, the receive signal $y(t)$ is defined as the time-delayed, phase-shifted, and attenuated signal $x(t)$ and AWGN $n(t)$:

$$y(t) = x(t) + n(t) \tag{1}$$

In case of a ULA, the individual sensor's receive signal is calculated using the steering vector $\mathbf{a}$, a vector that induces phase shifts on the $M$ sensors depending on the geometry of the array antenna and the angle of arrival:

$$\mathbf{y}(t) = \mathbf{a}(\theta)x(t) + \mathbf{n}(t), \quad \mathbf{y}(t) = [y_1(t), \ldots, y_M(t)]^T \tag{2}$$

If multipath is added, the signal $x(t)$ from (1) describes the superposition of all $K$ incoming waves. The simulations of multipath interference are based on the WINNER II channel model [9]. Its features include 17 distinct multipath propagation scenarios, various antenna configurations, Doppler shift, and wideband (up to 100 MHz) multiple-input multiple-output (MIMO) modeling. Channel parameters for each snapshot are calculated from statistical distributions based on channel measurements mainly obtained at carrier frequencies at 2 and 5 GHz. The modeled parameters are therefore not optimized for the E-GSM 900 band, but a comparison between the TDOA and AOA method can still be conclusive. In this paper, the scenario B1 "Urban micro cell" is used, which allows the distinction between multipath with and without a LOS signal.

# 3 Time difference of arrival estimation

The method time difference of arrival (TDOA) refers to the difference in time-of-arrival of a mobile phone signal at two spatially separated receivers:

$$\Delta\tau_i = \tau_1 - \tau_i, \quad i = 2, \ldots, n \tag{3}$$

with $n$ indicating the number of available base stations. Its equivalent distance is the product of the TDOA and the propagation speed (approximately the speed of light $c$):

$$\Delta d_i = \Delta\tau_i \cdot c, \quad i = 2, \ldots, n \tag{4}$$

The resulting range differences yield hyperbolic curves between base stations (BS), as depicted in Fig. 3. A mobile station's position (MS) lies at the intersection of these hyperbola. Assuming the positions of the base stations are known, multilateration can be performed with a minimum of three time-synchronized base stations (four in 3D).

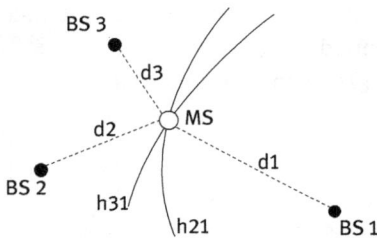

**Fig. 3.** Multilateration in TDOA.

In Section 2.2 it was pointed out that the temporal resolution of TDOA estimation is constrained by the signal bandwidth. Therefore, a bandwidth expansion technique is introduced to improve resolution in narrowband GSM. The SNR, however, is fixed by the individual measurement equipment and the observation length can only be maximized by feeding the estimator with full GSM bursts. Another effect may result from multipath signals, which interfere by inducing a measurement bias depending on their signal strength. Furthermore, inaccuracies in the time-synchronization process or positioning of the base stations need to be minimized as they degrade the performance of the whole system.

In this paper, the maximum likelihood estimator (MLE) is used for calculating the time differences $\Delta\tau$. The parametric estimator performs the probability density function (PDF) of a normal distribution on a set of noisy or incomplete observations. It is stated that the MLE has optimal asymptotic properties of being unbiased, and achieving the Cramer-Rao lower bound (CRLB) for an infinite number of samples [10].

The signal model for TDOA with two base stations is specified as the receive signal at BS 2 as a function of the receive signal at BS 1:

$$y_2(t) = \underbrace{y_1(t - \Delta\tau) \cdot \alpha}_{x_2(t)} + n(t), \tag{5}$$

where $\Delta\tau$ refers to the time difference, $\alpha$ represents the complex attenuation, and $n(t)$ AWGN. The receive signal $y_2(t)$ is considered an approximately random process with a Gaussian PDF. The signal model indicates a single source estimation, but may be modified if further clutter cancellation is required.

The likelihood function is defined as the joint PDF of the signal model for the unknown parameter $\theta = [\Delta\tau, \alpha, \sigma_n^2]^T$ and given observation intervals $y_2(N)$. $\sigma_n^2$ describes the noise variance and $N$ the number of independent discrete observations. In this particular case, it can be reduced to a function only dependent on $\Delta\tau$. The so-called reduced log-likelihood function reads

$$\Lambda = \frac{y_1(\Delta\tau)y_2{}^H y_2 y_1(\Delta\tau){}^H}{y_1(\Delta\tau)y_1(\Delta\tau){}^H}. \tag{6}$$

The maximum likelihood estimator $\hat{\Delta\tau}$ is the value that maximizes the reduced log-likelihood function and is found iteratively, where

$$\hat{\Delta\tau} = \arg\max_{\Delta\tau}(\Lambda). \tag{7}$$

Due to the analog to digital conversion, the MLE accuracy is bounded by the sampling interval. Therefore, it is recommended to perform additional polynomial interpolation subsequent to the estimation process. Simulations in this paper are performed with a 2nd order polynomial interpolation that approximates the maximum by fitting a parabola on three given data points:

$$\hat{\Delta\tau}_{max} = \frac{-2\Delta\hat{\tau}_0\Lambda_+ + \Lambda_+\Delta t + 4\Delta\hat{\tau}_0\Lambda_0 - 2\Delta\hat{\tau}_0\Lambda_- - \Lambda_-\Delta t}{-2\Lambda_+ + 4\Lambda_0 - 2\Lambda_-} \tag{8}$$

$\Delta\hat{\tau}_0$ is the MLE, $\Delta t$ the sampling interval, and $\Lambda_-$, $\Lambda_0$ and $\Lambda_+$ the values of the reduced log-likelihood function for the three given data points.

# 4 Angle of arrival estimation

Another method for mobile phone localization is based on determining the angle of arrival (AOA) of a signal. Antenna arrays at the base stations measure the difference in phase (time) of the signal between antenna elements. The phase difference is then converted into an angle. Their relation can be expressed as

$$\Delta\varphi_{io} = \frac{2\pi d_i(\theta)}{\lambda},$$
(9)

where $\Delta\varphi_{io}$ refers to the phase difference between the sensor element $i$ and the coordinate origin $o$. The distance $d_i$ is a function of the angle of arrival $\theta$, and $\lambda$ denotes the wavelength of the signal. As illustrated in Fig. 4, positioning of the mobile station is performed with a minimum of two base stations via triangulation (3D additionally requires the elevation angle). The positions of the base stations are assumed to be known.

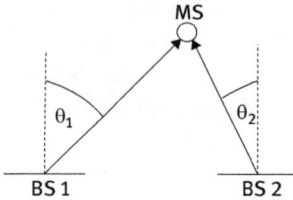

**Fig. 4.** Triangulation in AOA.

Simulations in this paper are conducted with a standard ULA antenna. A ULA consists of $M$ identical sensors that are distributed equidistant on a line. The term standard is used if the spacing between array elements is half the wavelength of the signal. A magnetometer is used to determine the antenna array's orientation. Errors in orientation measurement are directly transferred into the estimation error. In this paper, it is assumed that the orientation error is negligible. Also, antenna arrays need to be carefully calibrated to reduce mutual coupling effects as well as placed in the far-field of the source. Their accuracy may drop significantly for noisy nearly planar incoming waves [11].

Using the MLE for AOA estimation would mean a multi-dimensional (computational intensive) search. That is why subspace-based methods, which are significantly less computational complex, are employed. However, these estimators are declared suboptimal because they only approximate the CRLB for high SNR. Subspace methods for AOA estimation are (unlike TDOA) designed to detect narrowband signals. Hence, a single 200 kHz GSM burst works well as signal source. The SNR and the observation length have a similar effect on the accuracy as stated for TDOA. Multipath signals induce a measurement bias in case the angle of arrival differs from the direction of the LOS signal.

The subspace estimation methods follow a covariance matrix model. The receive signal is given as

$$\mathbf{y}(t) = \mathbf{A} \cdot x(t) + \mathbf{n}(t), \tag{10}$$

where $\mathbf{y}(t)$ is a $M$-long column vector of the antenna element outputs. The array manifold matrix $\mathbf{A} = [\mathbf{a}(\theta_1), \dots, \mathbf{a}(\theta_K)]$ holds all information on the geometry of the antenna array for a number of incoming waves $(K)$. Since $\mathbf{n}(t)$ is AWGN, the covariance matrix of $\mathbf{y}(t)$ becomes

$$\mathbf{R} = \mathbf{A}\mathbf{X}\mathbf{A}^H + \sigma_n^2 \mathbf{I}. \tag{11}$$

$\mathbf{X}$ denotes a nonsingular covariance matrix of the noise-free signal $x(t)$ [7]. For coherent signals, $\mathbf{X}$ is singular and a forward-backward spatial smoothing decorrelation becomes necessary [12].

Root multiple signal classification (Root-MUSIC) is a parametric estimator with higher resolution than the original MUSIC, but only valid for ULAs. In a first step, it computes the sample covariance matrix, on which it performs an eigendecomposition in order to obtain the noise subspace. Root-MUSIC is different from MUSIC as it estimates the AOA by determining the argument (phase) of the roots nearest the unit circle of a spectrum polynomial. The polynomial equation is based on the noise subspace. If the AOA is defined perpendicular to the ULA, $\hat{\theta}$ is obtained by

$$\hat{\theta}_i = \arcsin\left(\frac{c\Delta\varphi_i}{2\pi f_c \Delta a}\right), \tag{12}$$

where $\Delta\varphi_i$ denotes phase angle of the ith root nearest the unit circle [13]. As mentioned before, the distance between antenna elements in a standard ULA is $\Delta a = \lambda/2$. $f_c$ is the carrier frequency of the signal.

The second estimator proposed is the estimation of signal parameters via rotational invariance techniques (ESPRIT). It has the advantage of being far less computational intensive since it does not involve a search for maxima as in Root-MUSIC. ESPRIT divides the antenna array in two identical subarrays with a known displacement vector $\Delta\bar{a}$. This restricts the user to specific array configurations, e.g. the ULA. For ULAs, the displacement vector is the same as the interelement spacing $\Delta a$ [7].

Instead of using the noise subspace, ESPRIT determines the signal subspace of the twin subarrays $\mathbf{S}_1$ and $\mathbf{S}_2$. The ESPRIT requirement is formulated as

$$\mathbf{S}_2 = \mathbf{S}_1 \cdot \Phi, \tag{13}$$

where $\Phi$ describes a rotational matrix. The phase angle $\Delta\varphi_i$ is the ith eigenvalue of $\hat{\Phi}$, while $\hat{\Phi}$ is derived using Least-Square (LS) or Total-Least-Square (TLS) approximation. The AOA is calculated using (12).

It is noted that the subspace methods are limited to detecting a maximum of $M - 1$ signal sources. The estimation of the actual number of incoming waves (multipath signals) is critical as this number determines the dimension of the signal/ noise subspace. Its estimation is not included in the named methods and is usually done beforehand. Popular methods for this cause are the Akaike information criterion (AIC) or the minimum description length (MDL), which both originate from information theory [14]. Another method is based on sphericity tests, which in some cases yields better results than AIC or MDL [15]. In order to investigate the performance of the presented AOA estimators independently hereof, the number of signals here is fixed and assumed to be known.

# 5 Simulation results

The setup for the following simulations consists of two base stations located around a GSM mobile station, as sketched in Fig. 1. The distance to the mobile station is 80 and 100 m, respectively. Depending on the sampling frequency, the actual location varies by $\pm c/2f_s$ (here: $f_s$ = 40 MHz). It is possible to determine positioning accuracies by using techniques like the Geometric Dilution of Precision [16]. However, it is difficult to directly compare positioning results as the positioning method, number and placement of the base stations highly diversify the results. The following simulations solely investigate the capability of the localization methods TDOA and AOA for high-resolution detection of GSM signals.

## 5.1 TDOA estimation with LOS signals and AWGN

TDOA estimation is performed with the MLE and a 2nd order polynomial interpolation (see Section 3). The MLE search is stopped after a distance difference of 100 m as higher values may not be plausible (case-specific). The actual distance difference for the mentioned setup is 14.97 m, which gives a maximum error of about 85 m. The MLE is fed with a single GSM burst as well as with bandwidth expanded signals containing up to 25 bursts. In theory, a higher number of bursts (frequency channels) can be used, but the computational complexity rises linearly with the number of bursts.

Figure 5 illustrates the performance of the MLE in a LOS scenario without multipath by indicating the root-mean-square error (RMSE) for a range of SNRs. This figure outlines that TDOA estimation using a single GSM normal burst ($N$ = 1) is only reliable for a high SNR. As depicted in Fig. 5.b, there is a rapid increase in the magnitude of errors below a critical SNR of 27 dB. This can be explained by ambiguities in the peak detection of noisy signals when side-peaks approach the intensity of the main peak causing global errors. Side-peaks occur from signaling bits or the finite length of a signal. Eventually, below an SNR of −5 dB the AWGN superimposes with the likelihood of the estimator. The estimated difference in propagation distance of about 42 m is exactly the amount of variance one would expect from a wild guess.

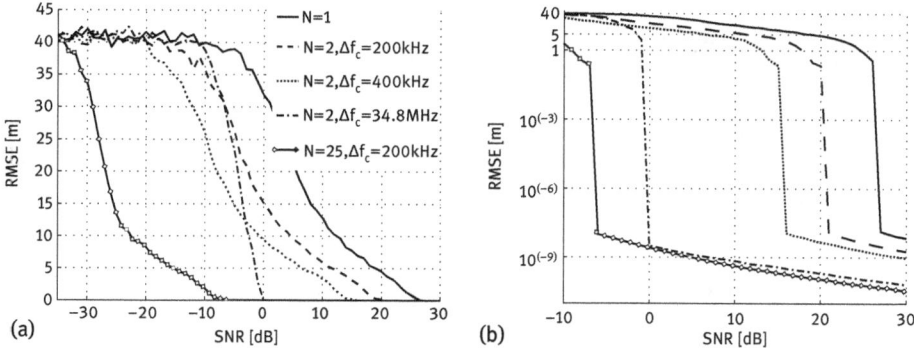

**Fig. 5.** RMSE of TDOA estimation with noisy LOS signals, 5000 simulation runs.

The figure further shows that the resolution can be enhanced by combining multiple bursts sent on different frequencies. A bandwidth expanded signal of two bursts with a frequency spread of 200 kHz lowers the critical SNR by 5 dB. By spreading the two bursts on a wider frequency range ($\Delta f_c$ = 34.8 MHz is the maximum for E-GSM), it is possible to further lower the critical SNR before global errors occur. However, the transition between isolated ambiguities and a wild guess becomes very narrow due to stronger side peaks. Hence, the effective bandwidth can only be increased with an increase in the number of bursts. The upper end for the simulations presented here is the MLE using 25 bursts with a frequency spacing of 200 kHz. The algorithm is then well-capable of resolving the TDOA, indicated by an RMSE of less than one meter for an SNR of −8 dB or more. This concludes that the TDOA technique using the MLE proves to be robust against AWGN, assuming a wideband signal is provided.

## 5.2 TDOA estimation with multipath signals

The impact of multipath propagation on the estimator is demonstrated in Fig. 6 by plotting the reduced log-likelihood function of the MLE algorithm (without polynomial interpolation) for a 25-burst signal with $\Delta f_c$ = 200 kHz. Additionally, the graph marks the LOS and the non-LOS (NLOS) signals according to the individual received signal power. The actual distance difference of 14.97 m can be clearly associated with the LOS signal. In this particular example, the LOS signal's power is about 20 dB greater than the next NLOS signal. However, as described in Section 3, multipath signals may be of greater signal strength, which eventually results in more dominant peaks after the actual distance difference.

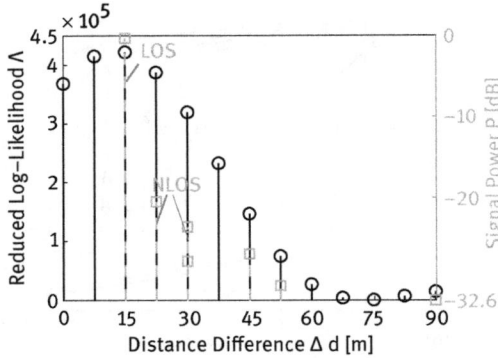

**Fig. 6.** Example of a reduced log-likelihood function with multipath.

TDOA simulations in the B1 multipath scenario (see Section 2.3) using the MLE with polynomial interpolation for varying bandwidths are listed in Tab. 1. The number of signals (LOS and NLOS) is set to $K = 7$. The results show that the superposition of the multiple signal sources lead to a considerable measurement bias, e.g. an RMSE of 11.4 m for a single burst in scenario B1 LOS. Furthermore and in contrast to the AWGN simulations, a larger frequency spread for two bursts degrades the accuracy due to more frequent ambiguities in the peak detection. However, if the bandwidth is sufficiently increased as with 25 bursts, the RMSE can be halved. Naturally, the bias is much larger in scenarios without LOS due to the longer journey of the NLOS signals. The RMSE for 25 bursts then reaches merely 20.6 m. Nevertheless, for TDOA it is possible to compensate parts of the bias effect resulting from multipath by conducting channel measurements matching the required application scenario.

## 5.3 AOA estimation with LOS signals and AWGN

AOA estimation is performed with the Root-MUSIC and ESPRIT algorithm as described in Section 4. In the mentioned setup, the distance between BS 1 and the mobile station is given as 97.30 m. The estimators performance is shown in Fig. 7 for a perpendicular incoming wave using a standard ULA of 2, 4, and 8 elements. The lower bound of the RMSE is denoted by the SQRT-CRLB. Due to the definition of the CRLB for local errors only, a comparison of the estimators to the CRLB fails below a certain SNR. Yet, the CRLB demonstrates that the accuracy can be increased with a higher number of array elements. Further, Root-MUSIC stands out as an efficient estimator by approaching the CRLB for SNR above −22 dB (8-element array). ESPRIT, on the other hand, is offset from the optimum for array configurations of more than 2 array elements, while the LS and the TLS-approximation yield similar results. Nevertheless, the results indicate that both algorithms are robust against noise. An RMSE of less than 1° is achieved with an SNR of −10 dB, even with a 2-element array.

**Tab. 1.** RMSE of TDOA estimation with multipath, 5000 simulation runs.

| Algorithm | Signal | B1 LOS | B1 NLOS |
|---|---|---|---|
| | $N = 1$ | 11.4 m | 23.3 m |
| | $N = 2, \Delta f_c = 200$ kHz | 12.0 m | 23.7 m |
| MLE | $N = 2, \Delta f_c = 400$ kHz | 12.8 m | 24.5 m |
| | $N = 2, \Delta f_c = 34.8$ MHz | 36.9 m | 38.4 m |
| | $N = 25, \Delta f_c = 200$ kHz | 5.2 m | 20.6 m |

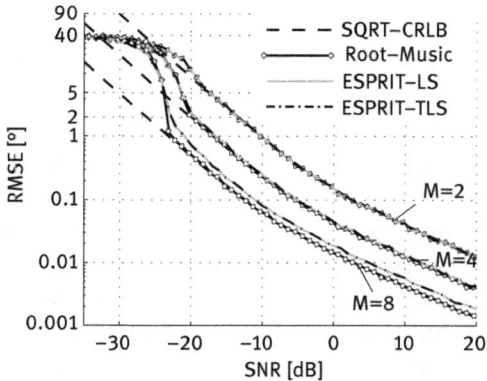

**Fig. 7.** RMSE of AOA estimation with noisy LOS signals, 500 simulation runs.

## 5.4 AOA estimation with multipath signals

The challenge coming from multipath propagation for AOA estimation is visualized in Fig. 8 with an example of a Root-MUSIC estimation on an 8-element antenna array and seven incoming signal sources. The figure indicates the absolute value (magnitude) of the polynomial roots on the y-axis and their associated AOAs on the x-axis. The signal power of the LOS and NLOS signals is depicted on the right hand side. In theory, the 8-element array is capable of detecting all seven paths, while the root nearest the unit circle is supposed to represent the LOS signal. In this particular example, the angle of the LOS signal is properly identified and strongly valued. However, the remaining paths are not recognized, and instead, the algorithm calculates roots at seemingly random angles. This leads to an estimated angle of 29 °, far off the actual AOA.

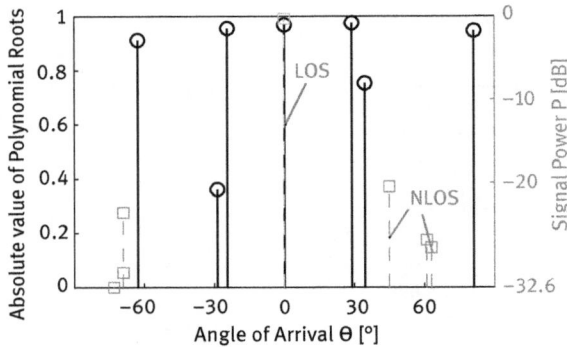

**Fig. 8.** Example of a Root-MUSIC estimation with multipath.

Simulations in the B1 multipath scenario ($K = 7$) using the Root-MUSIC and ES-PRIT algorithm with 2, 4, and 8 array-elements are summarized in Tab. 2.a. The results in the B1 NLOS scenario indicate that AOA estimation without a LOS to the mobile station is not feasible as NLOS signals arrive from all kinds of directions. Furthermore, it is shown that ESPRIT reaches higher accuracies for 4- and 8-element arrays than Root-MUSIC. The LS-approximation's performance is similar to the TLS-approximation. Surprisingly the best results are given for a 2-element array with an RMSE of 11.2 °. Subspace methods using 2 sensors are only able to detect a single path as the maximum number of resolvable paths is $M - 1$ (see Section 4). This suspects the AOA estimation process to show a better performance if only one path is resolved.

Therefore, the simulations are repeated in Tab. 2.b with the difference that only one path is set to resolve. The resulting angle can be interpreted as the average over all seven paths. Now, the estimators show a considerable smaller RMSE for 4 and 8 array elements. The RMSE of 7.2 ° and 7.1 ° of ESPRIT is about one degree smaller than that of Root-MUSIC. The simulations demonstrate subspace methods have severe problems with path allocation. Eventually, if only one path is resolved, a 4-element antenna array is recommended considering accuracy and cost of hardware.

**Tab. 2.** RMSE of AOA estimation with multipath, 5000 simulation runs.

(a)

| Algorithm | Array elements | B1 LOS | B1 NLOS |
|---|---|---|---|
| Root-MUSIC | $M = 2$ | 11.2 ° | 36.0 ° |
| | $M = 4$ | 34.6 ° | 46.9 ° |
| | $M = 8$ | 40.3 ° | 46.3 ° |
| ESPRIT-LS | $M = 2$ | 11.2 ° | 36.0 ° |
| | $M = 4$ | 24.5 ° | 35.1 ° |
| | $M = 8$ | 29.9 ° | 32.8 ° |
| ESPRIT-TLS | $M = 2$ | 11.2 ° | 36.0 ° |
| | $M = 4$ | 20.6 ° | 36.7 ° |
| | $M = 8$ | 29.9 ° | 33.8 ° |

(b)

| Algorithm | Array elements | B1 LOS | B1 NLOS |
|---|---|---|---|
| Root-MUSIC | $M = 2$ | 13.2° | 36.2° |
| | $M = 4$ | 8.2° | 35.7 ° |
| | $M = 8$ | 8.3° | 36.3 ° |
| ESPRIT-LS | $M = 2$ | 13.2° | 36.2 ° |
| | $M = 4$ | 7.2° | 33.8 ° |
| | $M = 8$ | 7.1° | 32.5 ° |
| ESPRIT-TLS | $M = 2$ | 13.2° | 36.2 ° |
| | $M = 4$ | 7.2° | 33.8 ° |
| | $M = 8$ | 7.1° | 32.5 ° |

# 6 Conclusion

The proposed methods TDOA and AOA are analyzed in simulations for high-resolution GSM mobile phone localization. First, it is shown to be essential to integrate the bandwidth expansion technique in the TDOA estimation process. This allows achieving a high resolution for the 200 kHz narrow GSM signals. Moreover, the MLE demonstrates the robustness of the method against noise. In the presence of multipath however, and especially if no line-of-sight (LOS) to the mobile station exists, the accuracy is reduced by an additional bias due to the superposition of multiple signal sources. If a LOS can be established, the RMSE reaches approximately 5 m for a bandwidth expanded signal of 25 bursts.

In comparison, the subspace methods Root-MUSIC and ESPRIT provide very accurate angle estimates for signals buried in noise. However, the simulations highlight that the AOA technique depends on a LOS to the mobile station and preferably little multipath, e.g. in open-space scenarios. A rough estimate of 7.2 ° for ESPRIT with 4 antenna elements can be obtained by resolving only a single signal source.

As search and rescue applications eventually face strong multipath signals, the TDOA method is recommended. TDOA is realized with a single antenna, but requires a minimum of three base stations and additional hardware for their time-synchronization. As presented, the LOS and WINNER II channel models differentiate the effects of channel noise and multipath interference. The I-LOV project further conducted channel measurements and initiated a channel model for collapsed buildings [17]. Based on this channel model, a sophisticated multilateration algorithm can be developed, which is expected to largely improve the TDOA system's performance with multipath.

**Acknowledgment:** This work has been created as part of the "I-LOV" project funded by the German Federal Ministry of Education and Research (BMBF).

# Bibliography

[1]  S. Zorn, R. Rose, A. Goetz, and R. Weigel. A novel technique for mobile phone localization for search and rescue applications. *Int. Indoor Positioning and Indoor Navigation (IPIN) Conf.*, 1–4, September 2010.

[2]  C. Drane, M. Macnaughtan, and C. Scott. Positioning GSM telephones. *IEEE Communications Magazine*, 36(4):46–54, 1998.

[3]  S. S. Wang, M. Green,and M. Malkawa. E-911 location standards and location commercial services. *IEEE Emerging Technologies Symp.: Broadband, Wireless Internet Access*, 2000.

[4]  K. J. Krizman, T. E. Biedka, and T. S. Rappaport. Wireless position location: fundamentals, implementation strategies, and sources of error. 47th *IEEE Vehicular Technology Conf.*, 2:919–923, May 1997.

[5]  *ETSI TS 45.001 Physical layer on the radio path*. Std., 3GPP.

[6]  A. N. Ekstrøm and H. J. Mikkelsen. *GSMsim A MATLAB Implementation of a GSM Simulation Platform*. Tech. rep., Institute of Electronic Systems, Aalborg University, Denmark, 1997.

[7]  P. Stoica and R. Moses. *Spectral Analyis of Signals*. Pearson/ Prentice Hall, 2005.

[8]  A. Goetz, R. Rose, S. Zorn, G. Fischer, and R. Weigel. A wideband crosscorrelation technique for high precision time delay estimation of frequency hopping GSM signals. 41st *European Microwave Conference*, 33–36, October 2011.

[9]  P. Kyösti, J. Meinilä, L. Hentilä, X. Zhao, T. Jämsä, C. Schneider, M. Narandzic, M. Milojevic, A. Hong, J. Ylitalo, V.-M. Holappa, M. Alatossava, R. Bultitude, Y. de Jong and T. Rautiainen. *WINNER II Channel Models ver 1.1*. Technical report, 2007. [Online]. Available: https://www.ist-winner.org/WINNER2-Deliverables/D1.1.2v1.1.pdf.

[10]  S. M. Kay. *Fundamentals of Statistical Signal Processing, Volume I: Estimation Theory*. Prentice Hall, 1993.

[11] A. L. Swindlehurst and T. Kailath. Passive direction-of-arrival and range estimation for near-field sources. 4th *Annual ASSP Workshop Spectrum Estimation and Modeling*, :123–128, August 1988.

[12] T. Shan, M. Wax, and T. Kailath. On spatial smoothing for direction-of-arrival estimation of coherent signals. *IEEE Trans. on Acoustics, Speech and Signal Processing*, 33(4):806–811, 1985.

[13] A. Barabell. Improving the resolution performance of eigenstructure-based direction-finding algorithms. *IEEE Int. Acoustics, Speech, and Signal Processing Conf. ICASSP'83*, 8:336–339, April 1983.

[14] M. Wax and T. Kailath. Detection of signals by information theoretic criteria. *IEEE Trans. on Acoustics, Speech and Signal Processing*, 33(2):387–392, 1985.

[15] D. B. Williams and D. H. Johnson. Using the sphericity test for source detection with narrow-band passive arrays. *IEEE Trans. on Acoustics, Speech and Signal Processing*, 38(11):2008–2014, 1990.

[16] N. Levanon. Lowest GDOP in 2-D scenarios. *IEE Proc. -Radar, Sonar and Navigation*, 147(3):149–155, 2000.

[17] L. Chen, M. Loschonsky, and L.M. Reindl. Autoregressive Modeling of Mobile Radio Propagation Channel in Building Ruins. *IEEE Trans. on Microwave Theory and Techniques*, 60(5):1478–1489, 2012.

# Biographies

**Lars Zimmermann** received the Dipl-Ing. degree in mechatronics from the University of Erlangen-Nuremberg, Germany, in 2010. From 2010 to 2013, he was a project leader and later funding projects coordinator at eesy-id GmbH, Erlangen, Germany. Currently, he is working toward the Ph.D. degree at the University of Erlangen-Nuremberg. His research interests include sensor fusion, MEMS, indoor air quality sensors, wireless communication, and radar.

**Alexander Goetz** has studied Electrical, Electronics, and Information Technology at Friedrich-Alexander-University Erlangen-Nuremberg, Germany, and graduated as Dipl.-Ing. Univ. in May 2008. Consequently, he has been engaged as research assistant and doctoral candidate at the Institute for Electronics Engineering at Friedrich-Alexander-University Erlangen-Nuremberg and graduated as Dr.-Ing. in September 2012. His research interests are in the fields of communication engineering, localization and radar technology and digital signal processing.

**Georg Fischer** received the Dr.-Ing. degree in electrical engineering from the University of Paderborn, Germany, in 1997. From 1996 to 2008, he performed research with Bell Laboratories, Lucent (later Alcatel- Lucent). In 2000, he became a Bell Labs Distinguished Member of Technical Staff, and in 2001, a Bell Labs Consulting Member of Technical Staff. He was also a Chairman with the European Telecommunications Standards Institute (ETSI) during the physical layer standardization of GSM-EDGE. From 2001 to 2007, he was a Part-Time Lecturer for base station technology, and since April 2008 a Professor of electronics engineering at the University of Erlangen-Nuremberg. Dr. Fischer holds over 50 patents concerning microwave and communications technology. He is a Senior Member of the IEEE Microwave Theory and Techniques Society (MTT-S), Antennas and Propagation Society (AP-S), Computer Society (COMSOC) and Vehicular Technology Society (VTC). He is a member of VDE-ITG and the European Microwave Association (EUMA).

**Robert Weigel** has been Director of the Institute for Communications and Information Engineering at the University of Linz, Austria during 1996 to 2002. In Linz, in 1999, he co-founded the company DICE, meanwhile split into an Infineon Technologies (DICE) and an Intel (DMCE) company with about 400 co-workers. Since 2002 he is Head of the Institute for Electronics Engineering at the University of Erlangen-Nuremberg, Germany. There, respectively in 2009, and in 2012 he co-founded the companies eesy-id and eesy-ic. Dr. Weigel has published more than 800 papers and received the 2002 VDE ITG-Award, the 2007 IEEE Microwave Applications Award and the 2016 IEEE MTT-S Distinguished Educator Award. He is a Fellow of the IEEE, an Elected Member of the German National Academy of Science and Engineering (acatech), and an Elected Member of the Senate of the German Research Foundation (DFG). He has been the 2014 MTT-S President.

R. Rahimi and G. Dadashzadeh

# Evaluation of Auxiliary Tone Based MAC Scheme for Wireless ad hoc Networks with Directional Antennas

**Abstract:** This paper presents an analytical evaluation of auxiliary tone based MAC scheme that has been proposed before to mitigate the directional antenna utilization problems in wireless ad hoc networks. We conduct a comparison study between two different Markov models for auxiliary tone based MAC scheme. In the current proposed model in addition to the idle, success, and fail states which has been suggested as different nodes' states during network communications, also the defer state has been considered which makes the model more realistic. Results show that based on this model the previously proposed scheme outperforms the RTS/CTS based MAC scheme not only in high density networks but also in networks with low probability of transmission and low density networks.

**Keywords:** ATB-DMAC, Directional antennas, Medium access control, Ad Hoc networks.

# 1 Introduction

One of the most important issues in wireless ad hoc networks is how to exploit directional antenna benefits while avoiding the MAC layer problems like directional hidden node, directional exposed node, and the deafness problems [1]. Various MAC schemes have been proposed to make a satisfactory trade off between benefits and drawbacks of utilizing directional antennas [8–19]. In [2] Deng et al. introduced Dual Busy Tone Multiple Access (DBTMA) which uses busy tones to prevent the omni-directional hidden node and deafness problems. In [3] the DBTMA has been extended to the nodes equipped with directional antennas for transmitting the busy tones. In [1] the deafness problem has been introduced and a busy tone based scheme has been proposed to mitigate it by informing the neighbors about end of the data transmission. The auxiliary tone based directional MAC scheme (ATB-DMAC) [4] is a useful MAC scheme that has been proposed to exploit the benefits provided by directional antennas in wireless ad hoc networks while attempting to mitigate the hidden node, exposed node,

___
**R. Rahimi and G. Dadashzadeh:** Electrical and Electronic Engineering Department, Shahed University, Tehran, Iran, email: rezgar_rahimi@yahoo.com.

De Gruyter Oldenbourg, ASSD – Advances in Systems, Signals and Devices, Volume 4, 2017, pp. 35–54.
DOI 10.1515/9783110448399-003

and deafness problems. One of the essential characteristic of ATB-DMAC scheme is its tone based neighbor discovery at the beginning of each packet transmission that can make it more suitable for mobile ad hoc networks with random movements. In [4], performance of this scheme has been evaluated based on the three states discrete Markov model in which ATB-DMAC outperforms RTS/CTS based directional MAC only in dense networks and in networks with high probability of transmission. In this paper we conducted a comparison evaluation based on the Markov model with four states in which the idle, success, fail, and defer states have been considered and show through analysis how ATB-DMAC can help to increase throughput also in networks with low probability of transmission and low number of network nodes. The rest of the paper is outlined as follows. Section 2 reviews the ATB-DMAC scheme. In section 3 the performance analysis is presented. Section 4 includes the numerical results. We conclude the paper in section 5.

## 2 ATB-DMAC scheme

As mentioned in [4], each node is equipped with an $M = 360/\theta$ (where $\theta$ is beam-width) elements switched beam antenna that can be used in Tone-Antenna (TA) and Packet-Antenna (PA) modes for transmitting tone and control/data packets, respectively. Adopting two-way ground model for transmission and signal-to-noise-plus-interference-ratio (SNIR) as successful reception criterion, we must have $SNIR = (P_t/P_k) = (R_k^4/R_t^4) \geq \sigma$. In which $\sigma = (R_i^4/R_t^4)$ is the SNIR threshold, $R_t$, $R_k$ and $R_i$ are transmission range, interferer-receiver distance and interference range, respectively. Each node with ATB-DMAC scheme uses following out-of-band tones to inform the neighbor nodes about its situation:
- Transmitter Direction Tone (TDT)
- Receiver Direction Tone (RDT)
- Other-direction Busy Tone (OBT)
- Same-direction Busy Tone (SBT)
- Desired Direction Tone (DDT)
- Collision Occurrence Tone (COT)
- RTS Collision Occurrence Tone (RCOT)

As noted before, nodes with ATB-DMAC scheme obtain the neighbor location information at the beginning of each packet transmission. Fig. 1 shows a comprehensive example that is set such that the application of each tone is considered. Suppose that in Fig. 1 (part 1) nodes A and F want to send data to nodes B and E, respectively. They start with TDT tone transmission on all M beams sequentially, to inform the neighbors about packet transmission and its own direction and also obtaining their neighbors' location information. Fig. 1 (part 2) shows that nodes B and G are idle and sense no tone but TDT tone, therefore they reply with the RDT tone to inform about their presence and their availability on that beam. Nodes C and D are communicating with each other, thus reply with OBT (to prevent the deafness by informing about involving in a connection in another direction, getting OBT means that no collision has happened and receiver did not reply because it was involved in another direction thus back-off time increasing is not required.) and SBT (to prevent the hidden node problem by informing about involving in a connection in the same direction, which prevents any future interferences) tones, respectively. Node E senses TDT tone on two different beams and replies with COT tone to inform the transmitters about simultaneous TDT transmission and to prevent probable collision occurrence on this direction. Thereafter, node F receives the COT tone and goes to the TDT defer state, which is designed to prevent future simultaneous TDT transmissions. Node A only replies with RTS packets on the directions that RDT tones are received and does not send any signal to the directions which the OBT, SBT, and COT tones have been received. Based on the destination address in RTS packet, since the received RTS packet is not related to the node G, it does not reply. In response to the successful RTS reception, node B transmits the DDT tone (to inform about intended node location) and CTS packet. Since the transmitter does not scan packet antenna for CTS packet but instead scans out of band frequencies using tone antenna, the DDT tone transmission decreases the waiting time. When transmitter detects the DDT tone, selects the related packet beam. Finally, proper reception of data packet in receiver is acknowledged by ACK packet transmission. Due to node movements it is possible to have simultaneous RTS reception. The RCOT tone is designed to inform about any RTS collision occurrence.

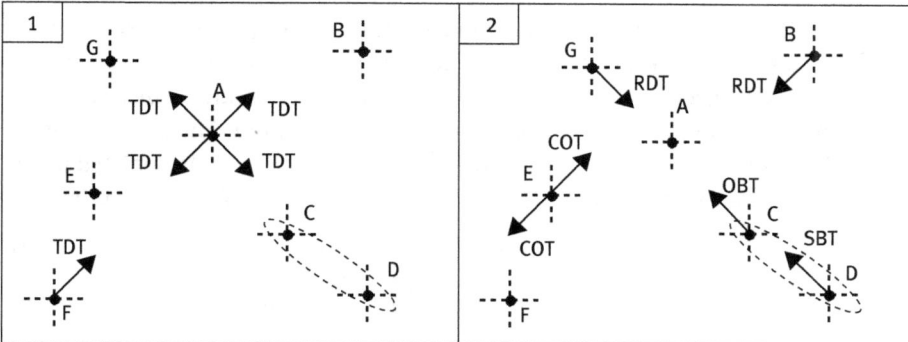

**Fig. 1.** Nodes' reaction in a network based on the ATB-DMAC scheme.

1. Four state Markov model for steady-state probabilities

2. Transmission and interference ranges [4]

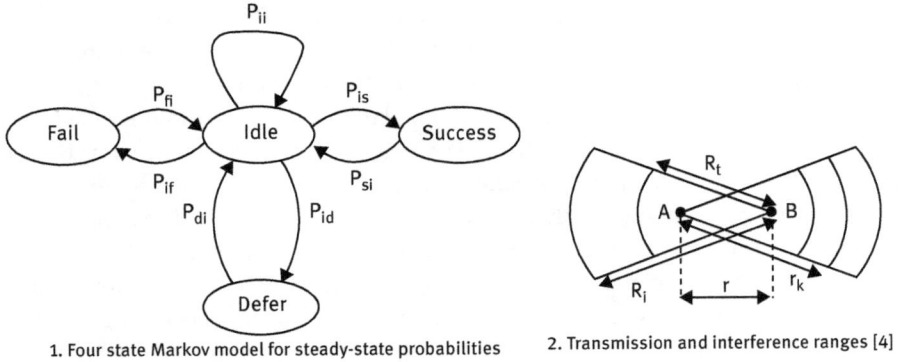

**Fig. 2.** Markov model and ranges for ATB-DMAC scheme.

# 3 Performance analysis

In [4], a three state discrete Markov model has been adopted for obtaining the saturation throughput. The proposed model in the current paper adopts the four state Markov model shown in Fig. 2 which has been used in [7] for performance evaluations. Based on this model saturation throughput and transmission delay can be calculated as follows:

$$Throughput = \frac{P(S)T_{DATA}}{P(I)T_i + P(S)T_s + P(D)T_d + P(F)T_f} \tag{1}$$

$$Delay = T_s + \frac{P(I)T_i + P(D)T_d + P(F)T_f}{P(S)} \tag{2}$$

where $P(I)$, $P(S)$, $P(D)$, and $P(F)$ are the steady-state probability of Idle, Success, Defer, and Fail states, and their duration has been denoted by $T_i$, $T_s$, $T_d$, and $T_f$, respectively. $T_{DATA}$ denotes the average data packet duration. Note that non-idle nodes will return to idle state after the corresponding duration with probability 1. Thus, assuming $P_{xy}$ as probability of transition from $x$ to $y$, we have: $P_{si} = P_{di} = P_{fi} = 1$. As depicted in Fig. 2 (part 1) each node is in one of the before-mentioned four states, which implies:

$$P(I) + P(S) + P(D) + P(F) = 1 \tag{3}$$

$$P(I) = P(I)P_{ii} + P(S)P_{si} + P(D)P_{di} + P(F)P_{fi} \tag{4}$$

Using (3) and (4), we obtain $P(I) = 1/(2 - P_{ii})$. Adopting Poisson distribution with density $\rho$ for node locations over two dimensional plane, the probability of finding $k$ nodes in an area of $A$ is equal to: $P(k, A) = \frac{(\rho A)k}{k!} \exp(-\rho A)$.

Let us denote the transmission probability of each node with $p$, the probability of transition from idle state to itself can be expressed as equation (5). From Fig. 2 (part 1) we can see that the steady-state probability of different states and the idle state are related via equations $P(S) = P(I)P_{is}$, $P(F) = P(I)P_{if}$, and $P(D) = P(I)P_{id}$.

$$P_{ii} = \sum_{k=0}^{\infty} \left(1 - p\frac{\theta}{2\pi}\right)^k \left(\frac{\rho A}{k!}\right) \exp(-\rho A) = \exp\left(-\rho p \frac{\theta}{2\pi} A\right) \qquad (5)$$

We define interference areas for different durations according to the depicted ranges in Fig. 2 (part 2), as below:

$$A_0 = 0, \ A_1 = R_t^2 \left(\frac{\theta}{2}\right)$$

$$A_2 = \left| R_t^2 \left(\frac{\theta}{2}\right) - \frac{1}{2}r^2 \tan\left(\frac{\theta}{2}\right) \right| \quad 0 \le r \le R_t \qquad (6)$$

$$A_3 = \left|
\begin{array}{ll}
0 & 0 \le r \le \dfrac{R_t}{\sqrt[4]{\sigma}} \\[2ex]
\sqrt{\sigma} r^2 \left(\dfrac{\theta}{2}\right) - R_t^2 \left(\dfrac{\theta}{2}\right) & \dfrac{R_t}{\sqrt[4]{\sigma}} \le r \le R_t
\end{array}
\right.$$

Let us denote the probability of successful reception in first time slot by $P_1$ and in TDT, RDT, RTS-CTS-ACK, DDT, DATA durations by $P_2$ to $P_6$, respectively. Hence, we have:

$$P_1 = \exp\left[-\rho p \left(\frac{\theta}{2}\right) A_1\right]$$

$$P_2 = P_3 = \exp\left[-\rho p \left(\frac{\theta}{2}\right) A_2(T_{Tone-Collision} + 2)\right]$$

$$P_4 = \exp\left[-\rho p \left(\frac{\theta}{2}\right) A_0(t_{RTS} + T_{CTS} + T_{ACK} + 3)\right] \qquad (7)$$

$$P_5 = \exp\left[-\rho p \left(\frac{\theta}{2}\right) A_3(T_{DDT} + 2)\right]$$

$$P_5 = \exp\left[-\rho p \left(\frac{\theta}{2}\right) A_0(T_{DATA} + 2)\right]$$

$$P_{is}(r) = \prod_{k=1}^{6} P_k$$

$$P_{is} = p(1 - p) \int P_{is}(r) f(r) dr, \ f(r) = 2r \qquad (8)$$

In which $T_{Tone-Collision}$ is the mean waiting time for tone transmission on each beam. Finally, we must obtain the idle, success, fail, and defer state durations. According to [5], we assume that $T_f$ follows a truncated geometric distribution:

$$T_f = \frac{1 - p}{1 - p^{T_2 - T_1 + 1}} \sum_{k=0}^{T_2 - T_1} p^k (T_1 + k)$$

$$T_1 = T_{TDT} + 1, \ T_2 = M(T_{TDT} + T_{RDT}) + t_{RTS} + 3 \qquad (9)$$

where $T_{TDT}$ and $T_{RDT}$ are time durations of TDT, RDT tones transmission, respectively. $T_{RTS}$ is time duration of RTS packet transmissions. The total RTS transmission time, can be derived as follows:

$$t_{RTS} = (MP_{inp})T_{RTS}, \quad P_{inp} = \sum_{k=0}^{\infty}((\rho A)^k/k!)exp(-\rho A) \tag{10}$$

In which $P_{inp}$ is the probability that at least one idle node presents in the transmission area $A$ of intended node. For simplicity, we assume $T_i = 1$ and:

$$T_s = T_d = T_{TDT} + T_{RDT} + t_{RTS} + T_{DDT} + T_{CTS} + T_{DATA} + T_{ACK} + 7 \tag{11}$$

Given that channel is idle, the transmission probability of each node in the next time slot have been obtained in [6]. A transmission can be successful or not. Also, when a node receives a packet from higher layer and senses the channel busy or does not receive any data from physical/higher layer it will not transmit. Hence:

$$P_{if} = p - P_{is}, \quad P_{id} = (1-p) - P_{ii} \tag{12}$$

For evaluating the ATB-DMAC scheme using new model we conducted throughput and delay comparison with the RTS/CTS based DMAC scheme proposed in [5]. Since in RTS/CTS based DMAC, nodes transmit all packets directionally and do not inform neighbor nodes before packet transmission, we must substitute $A_0$ with $A_3$, and in $P_2$ replace the $T_{Tone-Collision}$ with $T_{RTS}$. Also, $T_1 = T_{RTS} + 1$, and $T_2 = T_{RTS} + T_{CTS} + T_{DATA} + T_{ACK} + 4$.

# 4 Numerical results

In order to evaluate the above schemes we set parameters as follows: $\sigma = 10$, $\theta = n/8$, $T_{TDT} = T_{RDT} = T_{DDT} = 1$, $T_{RTS} = 20$, $T_{ACK} = T_{CTS} = 14$ and $T_{DATA} = 1024$. Fig. 3 (part 1) shows that when number of neighbor nodes is increased the ATB-DMAC outperforms the RTS/CTS based DMAC, e.g. in networks with $p = 0.15$, $p = 0.30$, and $p = 0.45$ for nodes with more than 20, 8, and 5 neighbor nodes, respectively. The results in Fig. 3 (part 2) show that, with four state Markov model, the ATB-DMAC scheme gives better throughput than the RTS/CTS based DMAC for wider range of network densities in comparison with the three state Markov model. Fig. 4 (part 3) shows that in a network with certain number of nodes the ATB-DMAC gives better performance for increasing in the probability of transmission. Also Fig. 4 (part 4) shows that based on the four and three state model throughput of nodes with 20 neighbor nodes that use the proposed scheme outperforms the RTS/CTS based scheme for $0.2 < p < 1$ and $0.5 < p < 1$, respectively, which means that the new scheme can outperform in wider range than that have been shown by the three state Markov model. Although, in part 2 and part 4 of Fig. 3 the three state model throughput is higher than that for the new model, but we shall notice that the defer state (also exponential back-off times [6] which has been considered in new model and can decrease the throughput severely) has not been considered in the previous model.

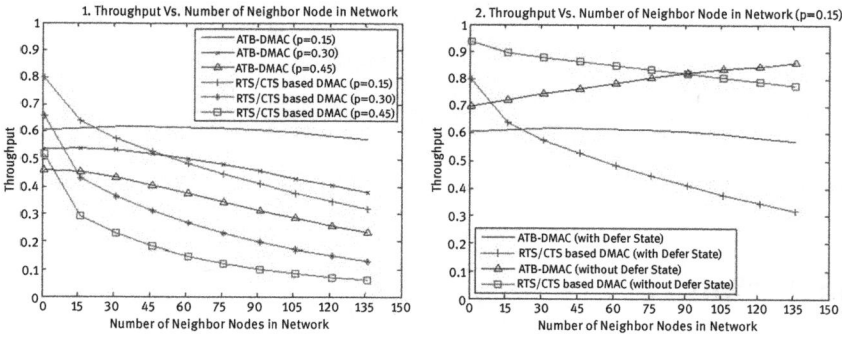

**Fig. 3.** Analytical comparison results: Throughput vs. number of network nodes.

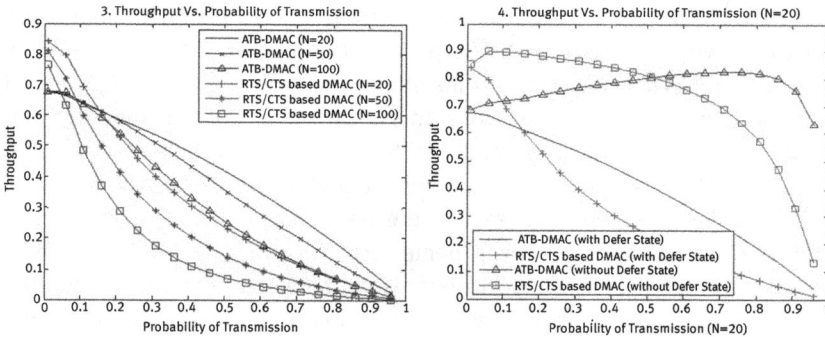

**Fig. 4.** Analytical comparison results: Throughput vs. probability of transmission.

As shown in Fig. 5 (part 1) due to neighbor discovery duration, with low number of neighbor nodes the proposed scheme's delay is higher than that for other scheme. For higher numbers, its delay increases slightly while it increases dramatically for latter scheme. Also, Fig. 5 (part 2) depicts that with same number of neighbor nodes for higher probability of transmission the proposed scheme's delay is much better than that for the RTS/CTS based scheme.

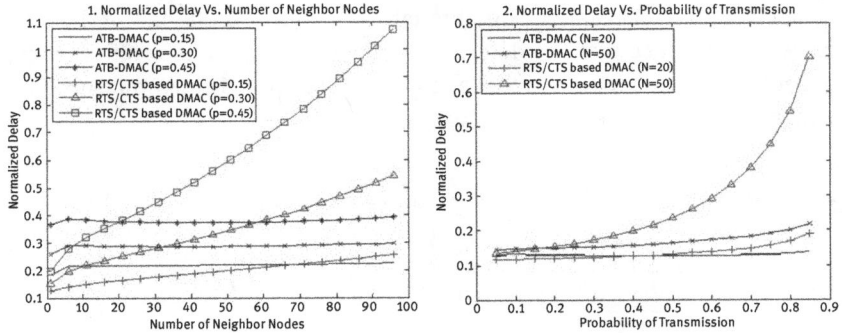

**Fig. 5.** Analytical comparison delay results.

# 5 Conclusion

In this paper we reviewed the ATB-DMAC scheme which has been proposed for mitigation of the problems that are introduced in the MAC layer of the ad hoc networks due to the utilization of directional antennas. As this scheme uses some extra auxiliary tones for neighbor discovery at the beginning of data communications, a small decrease in throughput and a small increase in transmission delay for short data or low density networks are expected. The previously proposed three state Markov model showed that this scheme outperforms other schemes such as RTS/CTS based MAC scheme only in high density networks. In the current performance evaluation, we proposed a more realistic Markov model with four states in which also the defer state is considered. The comparison evaluations are performed for the ATB-MAC and the RTS/CTS based MAC schemes based on the new analytical model. Numerical results show that the proposed ATB-MAC scheme outperforms the RTS/CTS based MAC scheme not only in high density networks but also in networks with low probability of transmission and in low density networks. This reasonable performance of the ATB-DMAC scheme is obtained due to the satisfactory utilization of the auxiliary tones in preventing the occurrence of long time defer states which is a notable parameter in decreasing the throughput and forcing long delays.

**Acknowledgment:** The authors express their gratitude to the anonymous reviewers for their thorough reviews and constructive comments.

This work is dedicated to my late father Seyed Hassan Rahimi.

# Bibliography

[1] R. R. Choudhury and N. H. Vaida. Deafness: A MAC problem in ad hoc networks when using directional antennas. *Int. Conf. on network protocols (ICNP)*, :283–292, Berlin, Germany, October 2004.

[2] J. Deng and Z. J. Haas. Dual busy tone multiple access (DBTMA): Performance evaluation. *IEEE Int. Veh. Technology Conf. (VTC)*, 1:314–319, Houston, TX, USA, May, 1999.

[3] C. S. Z. Huang, C.-C. Shen and C. Jaikaeo. A busy-tone based directional MAC protocol for ad hoc networks. *Military Communications Conf.*, 2:1233–1238, Anaheim, CA, USA, October 2002.

[4] R. Rahimi, G. Dadashzadeh, M. Maleki and E.Jedari. An auxiliary tone based MAC scheme for high density ad hoc networks with directional antenna. 15th *Asia-Pacific Conf. on Communications (APCC)*, Shanghai, China, October 2009.

[5] H. N. Dai, K. W. Ng, M. Y. Wu and B. Li. On collision-tolerant transmission with directional antennas. *Wireless Communication and Networking Conf. (WCNC)*, :1968–1973, Las Vegas, NV, USA, Mar. 2008.

[6] G. Bianchi. Performance analysis of the IEEE 802.11 distributed coordination function. *IEEE J. on Selected Areas in Communications*, 18(3):535–547, Mar. 2000.

[7] H. Ma, H. M. K. Alazemi and S. Roy. A stochastic model for optimizing physical carrier sensing and spatial reuse in wireless ad hoc networks. *IEEE Int. Conf. on Mobile ad-hoc and sensor systems (MASS'05)*, 8:615–622, Washington, DC, USA, Nov. 2005.

[8] J. Deng and Z.J. Haas. Dual busy tone multiple access (DBTMA): A new medium access control for packet radio networks. *IEEE Int. Conf. on Universal Personal Communicationsthe (ICUPC)*, 2:973–977, Florence, Italy, October 1998.

[9] V. Shankarkumar, Y.B. Ko and N.H. Vaidya. Medium access control protocols using directional antennas in ad hoc networks. *IEEE Computer and Communications Societies (INFOCOM)*, 1:13–21, Tel Aviv, Israel, March 2000.

[10] A. Nasipuri, S. Ye and R.E. Hiromoto. A MAC protocol for mobile ad hoc networks using directional antennas. *IEEE Wireless Communications and Networking Conf. (WCNC)*, 3:1214–1219, Chicago, IL, USA, Sep. 2000.

[11] M. Takai, J. Martin, A. Ren and R. Bagrodia. Directional virtual carrier sensing for directional antennas in mobile ad hoc networks. 3rd *ACM Int. Symp. on Mobile Ad Hoc Networking and Computing*, : 183–193, Lausanne, Switzerland, June 2002.

[12] R. R. Choudhury, X. Yang, R. Ramanthan and N. H. Vaidya. Using directional antennas for medium access control in ad hoc networks. *Int. Conf. on Mobile Computing and Networking (MobiCom)*, pp. 59-70, Atlanta, GA, USA, September 2002.

[13] T. Korakis, G. Jakllari and L.Tassiulas. A MAC protocol for full exploitation of directional Antenna in ad-hoc wireless networks. 4th *ACM Int. Symp. on Mobile Ad Hoc Networking and Computing*, :98–107, Annapolis, MD, USA, June 2003.

[14] H. Singh and S. Singh. Smart-802.11b MAC protocol for use with smart antennas. *IEEE Int. Conf. on Communication (ICC2004)*, 6:3684–3688, Paris, France, June 2004.

[15] R. Ramanthan. On the performance of ad hoc networks with beamforming antennas. 2nd *ACM Int. Symp. on Mobile Ad Hoc Networking and Computing*, :95–105, Long Beach, CA, USA, October 2001.

[16] N. S. Fahmi, T. D. Todd and V. Kezys. Ad hoc networks with smart antennas using IEEE 802.11-based protocol. *IEEE Int. Conf. on Communication (ICC2002)*, 5:3144–3148, Newyork, NY, USA, 2002.

[17] L. Bao, J.J. Garcia-Luna-Aceves. Transmission scheduling in ad hoc networks with directional antennas. *Int. Conf. on Mobile Computing and Networking (MobiCom)*, :48–58, Atlanta, GA, USA, September 2002.
[18] J. P. Monks, V. Bhargavan and W. W. Hwu. A power controlled multiple access protocol for wireless packet networks. *IEEE Computer and Communication Societies (INFOCOM)*, 1:219–228, Anchorage, AK, USA, April 2001.
[19] S-L Wu, Y-C Tseng and J-P Sheu. Intelligent medium access for mobile ad hoc networks with busy tones power control. *Int. Conf. of Computer Communications and Networking*, :71–76, Boston, MA, USA, October 1999.

# Biographies

**Razgar Rahimi** was born in Saghez, Kurdistan, Iran in 1981. He received the B.Sc. degree in electrical engineering from Iran University of Science and Technology, Tehran, Iran in 2005, and the M.Sc. degree in electrical engineering from Shahed University, Tehran, Iran in 2010. His recent research interests include cognitive radio networks and cooperative communication systems.

**Gholamreza Dadashzadeh** was born in Urmia, Iran, in 1964. He received the B.Sc. degree in communication engineering from Shiraz University, Shiraz, Iran, in 1992 and M.Sc. and Ph.D. degrees in communication engineering from Tarbiat Modarres University, Tehran, Iran, in 1996 and 2002, respectively. From 1998 to 2003, he has worked as the Head Researcher of the Smart Antenna for Mobile Communication Systems and the wireless local-area network 802.11 project with the radio communications group of Iran Telecommunication Research Center (ITRC). From 2004 to 2008, he was the Dean of the Communications Technology Institute, ITRC. He is currently an Assistant Professor with the Electrical and Electronic Engineering Department, Shahed University, Tehran. He has published more than 80 papers in referred journals and international conferences in the area of antenna design and smart antennas. Dr. Dadashzadeh is a member of the Institute of Electronics, Information, and Communication Engineers of Japan and the Iranian Association of Electrical and Electronics Engineers. He received the first degree of national researcher in 2007 from Iran's Ministry of Information and Communications Technology.

R. Rahimi and G. Dadashzadeh

# Improved Power Allocation in Parallel Poisson Channels

**Abstract:** The optimum power allocation for parallel Poisson channels which can minimize average power for obtaining maximum capacity is considered. In this paper a new power allocation scheme that outperforms previous schemes for more than enough powers is presented. Since in Poisson channels unbounded channel feeding with input power does not lead to a better utilization of the channel capacity, in the presented scheme extra powers that are more than required power for achieving maximum capacity are not used. If power dissipation is not possible, the extra power can be allocated in a new way which shows an improvement in the channel capacity. This approach is applied to the 2-fold parallel Poisson channels and is generalized to n-fold.

**Keywords:** Poisson Channels, Parallel channels, Optimum power allocation, Free-space optical communication, Capacity.

## 1 Introduction

Free Space Optics (FSO) is emerging as a popular technology because of its low-cost, high data rate communication and its several applications [4] and references therein. The Poisson channel has been accepted as a standard model for optical communication channels [7]. Shot noise, thermal noise, and laser intensity noise are well-known noises in optical intensity-modulated communications [1, 8]. As mentioned in a literature review of Poisson communication theory in [6], communication under Poisson regime was presented by I. Bar-David [9]. The capacity of Poisson channels under peak and average power constraints was proposed in [2, 3], and the capacity region of Poisson multiple-access channel is investigated in [7]. In [10, 11], the Poisson broadcast channels and the Poisson multi-input multi-output channels are studied, respectively. In free-space optical channels, atmospheric turbulence can lead to random fluctuation in intensity of optical signal and form a Poisson fading channel which is considered in [12]. Also, the information outage probability of a shot-noise limited direct detection MIMO optical channel subject to block fading is considered in [13] which has direct relevance to the results of [14, 15]. A new

**R. Rahimi and G. Dadashzadeh:** Electrical and Electronic Engineering Department, Shahed University, Tehran, Iran, email: rezgar_rahimi@yahoo.com.

De Gruyter Oldenbourg, ASSD – Advances in Systems, Signals and Devices, Volume 4, 2017, pp. 55–73.
DOI 10.1515/9783110448399-004

relationship between mutual information and conditional mean estimation in Poisson channels is found in [5]. A 2-fold and its generalized to n-fold parallel Poisson channel is considered in [1], in which a power allocation scenario that attempts to maximize the capacity of a peak and average power limited parallel Poisson channel is proposed. In the presented paper the above-mentioned power allocation for parallel Poisson channel is investigated and an improved scheme is proposed which shows that for powers that are more than enough, achievable capacities are higher than what has been stated in [1]. In this scheme despite the previous one, the extra power can be dissipated or allocated in a different way to get better results. The reminder of this paper is organized as follows. In section 2, a simple Poisson channel model and parallel Poisson channel models are outlined. Discussion about the alleged optimum power allocation and the new improved scheme are presented in section 3. Performance comparisons are provided in section 4, and conclusions are given in section 5.

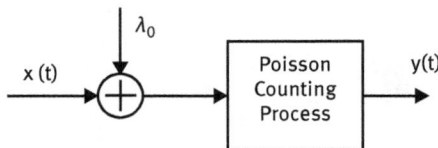

**Fig. 1.** Poisson channel model.

# 2  Parallel poisson channel

## 2.1  Poisson channel

In the Poisson channel model shown in Fig. 1, according to [7] for channel input $x(t) \geq 0$ and constant $\lambda_0 \geq 0$, which represents both the dark current and background noise, the channel output $y(t)$ is a doubly stochastic Poisson process with instantaneous rate $y(t) = x(t) + \lambda_0$, which is the number of photoelectrons counted in the interval $[0, T]$ by the direct detection device (photo-detector). Channels under peak and average limits are constrained to satisfy:

$$0 \leq x(t) \leq A, \qquad \frac{1}{T} \int_0^T x(\tau) d\tau \leq \sigma A \tag{1}$$

where the peak power $A > 0$ and the ratio of average to peak power $0 \le \sigma \le 1$ are constant. In [2, 3], the Shannon capacity of Poisson channel is given by:

$$C(A, \sigma, s) = A[p(1 + s) \ln(1 + s) + (1 - p)s \ln s - (p + s) \ln(p + s)] \qquad \text{nats/s} \qquad (2)$$

where $s = \dfrac{\lambda_0}{A}$ and $p = \min \left( \sigma, q(s) = \dfrac{(1 + s)^{1+s}}{s^s e} - s \right)$.

According to [1], the derivative of Poisson channel capacity as function of average-to-peak power ratio is given by $\dfrac{dC}{dp} = A \ln \left[ \dfrac{(1 + s)^{1+s}}{s^s e(1 + p)} \right]$. Therefore, regardless of average power constraint the maximum capacity is independent of $\sigma$ and is equal to:

$$C_{max} = A \left[ \frac{(1 + s)^{1+s}}{s^s e} - s(1 + s) \ln \left( 1 + \frac{1}{s} \right) \right] \qquad \text{nats/s} \qquad (3)$$

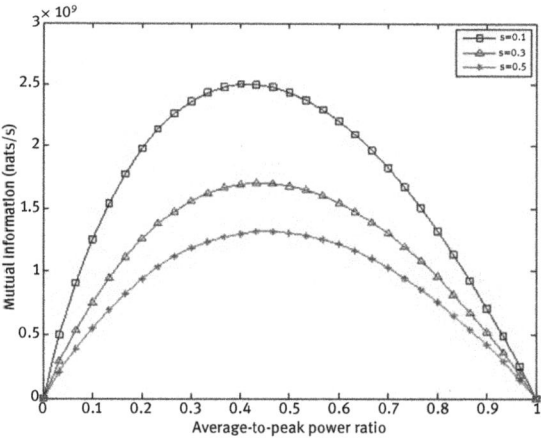

**Fig. 2.** Poisson mutual information versus average-to-peak power ratio for different signal-to-noise-ratios ($A = 10^{10}$).

As mentioned in [4], for OOK signaling scheme, the transmitted power has direct relationship to the duty cycle of the transmit aperture. Figure 2 illustrates the Poisson channel mutual information versus average-to-peak power ratio or duty cycle $\sigma$ for three different signal-to-noise-ratios ($SNR = 1/s$) which depicts three important points. First, It shows that with higher $SNR$ ($s = 0.1$), higher amount of maximum mutual information is achievable. Second, it shows that for fixed peak power $A = 10^{10}$ the maximum capacity for channels with lower signal-to-noise-ratio can be achieved with lower amount of channel input power. Finally, it reveals that by increasing $\sigma$ the Poisson channel capacity raises up to a maximum value and then declines.

## 2.2 Parallel Poisson channel

A parallel Poisson channel with n-independent channels can be modeled as n-independent single Poisson channels. Figure 3 shows an n-fold parallel Poisson channel such that peak and average input power of channel $i$ are restricted to $A_i$ and $\sigma_i A_i$, respectively. Under peak and average power constraints, each independent channel could use a certain amount of total input power. Therefore, the total channel capacity is the summation of individual channel capacities under above-mentioned restrictions:

$$C_{total} = \sum_i C_i \quad P_{total} = \sum_i \sigma_i A_i \tag{4}$$

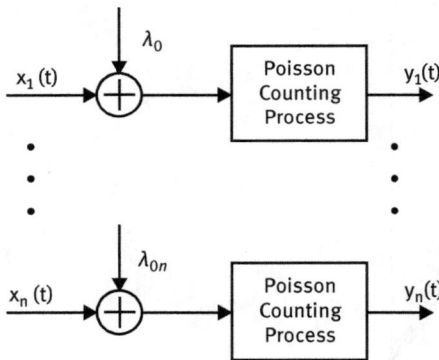

**Fig. 3.** Parallel Poisson channel.

Fig. 4 illustrates the influence of different power allocations schemes on total capacity of a two parallel Poisson channels with $s_1 = 0.20$, $s_2 = 0.05$, and $A_1 = A_2 = 10^9$. It shows that the maximum total rate $4.9163 \times 10^8$ is achievable for a certain case that the total input power is equal to the summation of required powers for obtaining maximum capacity on each individual channel. Therefore, for two channels with same peak power limits $A = A_1 = A_2$, we have: $\sigma_{tot-max} = \sigma_{1max} + \sigma_{2max}$, where $\sigma_{tot-max}$, $\sigma_{1max}$, and $\sigma_{2max}$ are optimum duty cycles for obtaining maximum capacity on two parallel channels, channel one, and channel two, respectively. Fig. 4 shows black and red curves that depict the individual channel capacities. As mentioned before Poisson channel with higher $SNR$ ($s_2 = 0.05$) needs lower power for attaining the maximum capacity. Note that like one single Poisson channel, the total capacity of

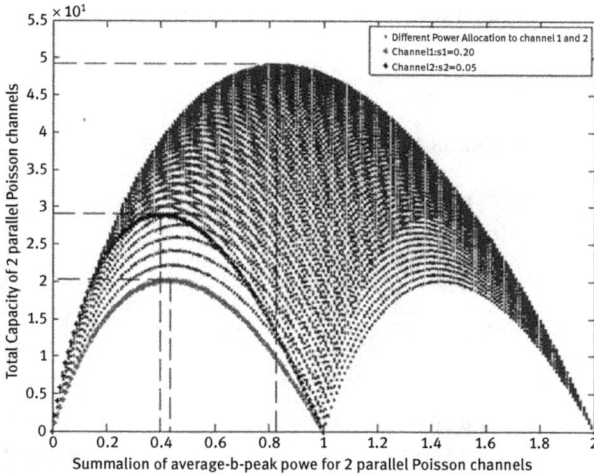

**Fig. 4.** Total capacity of two parallel Poisson channels for $s_1 = 0.20$ and $s_2 = 0.05$ and $A = 10^{10}$ with different power allocations.

parallel Poisson channel does not increase unboundedly by feeding higher amount of input power. These facts reveal the necessity of an optimum power allocation for improving the channel capacity.

# 3 Optimum power allocation

Applying different power allocation schemes can change the attainable capacity of parallel Poisson channels considerably. In [1], a power allocation scheme is introduced as optimum power allocation for these channels. In this section we investigate the proposed scheme and show that by modifying it for powers more than sufficient power for reaching the maximum total capacity, the channel utilization can be improved. For finding the optimum power allocation, following optimization problem should be solved:

$$\begin{aligned} \max \quad & C = C_{total}(\sigma_1, \sigma_2, ..., \sigma_n) = \sum_i C(\sigma_i) \\ \text{Subject to:} \quad & G(\sigma_1, \sigma_2, ..., \sigma_n) = \sum_i \sigma_i A_i = P_{total} \end{aligned} \tag{5}$$

We use maximization method of Lagrange multipliers to find best amounts of duty cycles. For simplicity, according to [1], at first we try to solve the problem for 2-fold parallel Poisson channel, and then generalize it to the n-fold case. Following this method, we must find the multiplier $\gamma$ such that $\nabla C = \gamma \nabla G$. We have:

$$\frac{\partial C}{\partial \sigma_1} = \gamma \frac{\partial G}{\partial \sigma_1} = \gamma A_1, \qquad \frac{\partial C}{\partial \sigma_2} = \gamma \frac{\partial G}{\partial \sigma_2} = \gamma A_2 \tag{6}$$

Solving above equations the following results are obtained [1]:

$$\sigma_1 = \left[ e^{-\gamma} q(s_1) - (1 - e^{-\gamma}) s_1 \right]^+$$
$$\sigma_2 = \left[ e^{-\gamma} q(s_2) - (1 - e^{-\gamma}) s_2 \right]^+ \tag{7}$$

where $(x)^+ = max(x, 0)$, and:

$$e^{-\gamma} = \frac{s_1 A_1 + s_2 A_2 + P}{[s_1 + q(s_1)]A_1 + [s_2 + q(s_2)]A_2} \tag{8}$$

From equation (7) and considering Fig. 4 as an example we may have three different conditions. First, when one of the duty cycle equations $[e^{-\gamma} q(s_i) - (1 - e^{-\gamma}) s_i]$ become lower than zero. In this case as equation (7) and Fig. 4 suggest we must dedicate all input power to the other channel. Second is the condition that none of duty cycles are lower than zero, and the total power is smaller than enough power for reaching the maximum total capacity which is $P_{enough} = q(s_1))A_1 + q(s_2)A_2$. In this case as suggested in [1], we allocate the available power to the channels according to equation (7). The third condition which is related to the main contribution of this paper is when the total available power is more than enough power $P_{enough}$. In this condition unlike the suggested approach in [1] we have two different approaches; ifallocation of total power is not mandatory, we can highly improve the channel capacity with dissipating the non-required power which is more than enough that declines the total channel capacity. Utilizing this approach keeps the capacity in the maximum value. Therefore, for $P_{total} > q(s_1)A_1 + q(s_2)A_2$ we suggest $\sigma_i = q(s_i)$. If it is not possible to dissipate this extra power there is another power allocation scheme which although has not significant increase in capacity but shows that the alleged optimum power allocation in parallel Poisson channel can be improved slightly. In this case proposed values for duty cycles are same as equation (7). For n-fold parallel Poisson channel this approach can be simply generalized. For these channels if first condition happens, channels with negative values for duty cycle are excluded and the total power could be dedicated to the remained channels. For second and third conditions, the approach is similar to that for 2-fold case. Supposing an n-fold parallel Poisson channel, the abovementioned approach is summarized as follows:

- If we have $[e^{-\gamma} q(s_i) - (1 - e^{-\gamma}) s_i] \le 0$, for $i \in L$, $\|L\| < n - 1$, we set $\sigma_i = 0$ and for remained channels that $[e^{-\gamma} q(s_i) - (1 - e^{-\gamma}) s_i] > 0$, we allocate total power such that only $n - \|L\|$ channels exist. Therefore we have:

$$\begin{cases} \sigma_i = 0 & i \in L \\ \sigma_i = [e^{-\gamma} q(s_i) - (1 - e^{-\gamma}) s_i]^+ & i \notin L \end{cases} \tag{9}$$

where:

$$e^{-\gamma} = \frac{\sum\limits_{i \notin L} s_i A_i + P}{\sum\limits_{i \notin L} [s_i + q(s_i)] A_i} \tag{10}$$

– If we have $i \in L$, $\|L\| = n - 1$, such that $[e^{-\gamma} q(s_i) - (1 - e^{-\gamma}) s_i] \leq 0$, we set $\sigma_i = 0$ and forthe remained channel that $[e^{-\gamma} q(s_i) - (1 - e^{-\gamma}) s_i] > 0$, we allocate total power to this channel. Therefore we have:

$$\begin{cases} \sigma_i = 0 & i \in L \\ \sigma_i = \dfrac{P_{total}}{A_i} & i \notin L \end{cases} \tag{11}$$

– If $P_{total} \geq q(s_1) A_1 + q(s_2) A_2$ and power dissipation is allowed we set $\sigma_i = q(s_i)$
– If $P_{total} \geq q(s_1) A_1 + q(s_2) A_2$ and total available power allocation is mandatory, we set $\sigma_i = (e^{-\gamma} q(s_i) - (1 - e^{-\gamma}) s_i)^+$. Where:

$$e^{-\gamma} = \frac{\sum\limits_{i} s_i A_i + P}{\sum\limits_{i} [s_i + q(s_i)] A_i} \tag{12}$$

# 4 Performance comparison

In this section, we analytically compare the performance of alleged optimum power allocation [1], with performance of two new proposed schemes for powers which are more than enough for obtaining maximum capacity. For proper comparison, values are same as those used in [1]. Therefore, we have a 2-fold parallel Poisson channel with $A_1 = 10^9$, $A_2 = 10^{12}$, $s_1 = 0.1$, and $s_2 = 0.3$.

Figure 5 illustrates three different count intensity allocations to individual channels versus input average count intensity. It can be observed that the alleged optimum scheme and the new scheme without power dissipation allocate all the input power to the individual channels, while in the third scheme which power dissipation is allowed only the enough power is allocated to the individual channels and the remained power is not used. Also, in comparison between new scheme without power dissipation and the referred one [1], it can be seen that the portion of input average count intensity that is dedicated to the stronger channel ($s_1 = 0.1$, $A_1 = 10^9$) in first scheme is lower than that for latter one. Instead this power is assigned to the weak channel ($s_2 = 0.3$, $A_2 = 10^{12}$). Fig. 6 depicts the effect of these three different power allocation schemes on channel rate, in which the parallel Poisson channel rate for average input powers more than optimum power are presented. As expected the channel rate remained fixed with the scheme that dissipates the extra power, and a closer observation of

Fig. 6 indicates that the new scheme which does not dissipate the extra power but dedicates it in a different way, outperforms the alleged optimum power allocation scheme slightly.

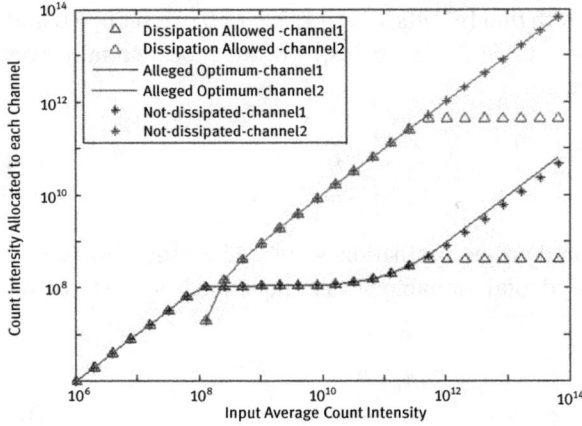

**Fig. 5.** Comparison among three different count intensity allocation schemes with parameters $A_1 = 10^9$, $A_2 = 10^{12}$ photons/s, $s_1 = 0.1$ and $s_2 = 0.3$.

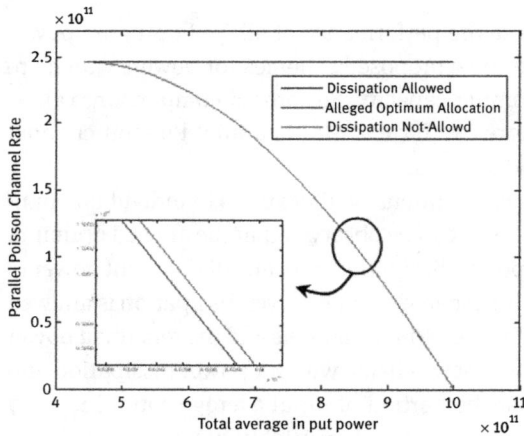

**Fig. 6.** Parallel Poisson channel rate comparison for three different power allocation schemes.

# 5 Conclusion

In this paper, an improved power allocation scheme with peak and average power constraints for parallel Poisson channel is considered and it is shown that this scheme with two different approaches can obtain more channel rates than that for previous alleged optimum power allocation. Our proposed schemes and the previous scheme follow the same approaches when the available power is les than the required power for reaching the maximum available rate which is obtained through Lagrangian multipliers. The main difference occurs for input average powers that are more than the required power for reaching the maximum channel capacity. For this situation we proposed two different approaches, the first one is for transmitters that can dissipate the extra powers. The second one is for situations that all available input power shall be used for transmission, which based on the proposed scheme even for this case, we can obtain higher rates over channels in comparison to the allegedly optimal power allocation scheme. Considering the performance results depending on the dissipation possibility, these approaches can substitute the previous schemes for power allocation in parallel Poisson channels.

**Acknowledgment:** The authors express their gratitude to the anonymous reviewers for their thorough reviews and constructive comments.

This work is dedicated to my late brother-in-law Anvar Davari.

# Bibliography

[1] M. M. Alem-Karladani, L. Sepahi, M. Jazayerifar and K. Kalbasi. Optimum power allocation in parallel Poisson optical channel. *Int. Conf. on Telecommunications*, Zagreb, 2009.

[2] Y. M. Kabanov. The capacity of a channel of the Poisson type. *Theory of Probability & its Applications*, 23:143–147, 1978.

[3] M. H. A. Davis. Capacity and cut-off rate for Poisson-type channels. *IEEE Trans. Information Theory*, IT-26(6):710–715, November 1980.

[4] K. Chakraborty, and S. Dey and M. Franceschetti. Service-outage-based power and rate control for Poisson fading channels. *IEEE Trans. Information Theory.*, 55(5):2304–2318, May 2009.

[5] D. Guo, S. Shamai and S. Verdu. Mutual information and conditional mean estimation in Poisson channels. *IEEE Trans. Information Theory*, 54(5):2304–2318, May 2008.

[6] S. Verdu. Poisson communication theory. *Int. Technion Commuincation Day in Honor of Israel Bar-David*, Haifa, Israel, March 1999.

[7] A. Lapidoth and S. Shamai. The Poisson multiple-access channel. *IEEE Trans. Information Theory*, 44(2):488–501, March 1998.

[8] R. M. Gagliardi and S. Karp. *Optical communications*. John Wiley and Sons, New York, 2nd ed., 1995.

[9] I. Bar-David. Communication under the Poisson regime. *IEEE Trans. Information Theory*, 15:31–37, 1969.

[10] A. Lapidoth, I. E. Telatar and R. Urbanke. On wide-band broadcast channels. *IEEE Trans. Information Theory*, 49(12):3250–3258,December 2003.

[11] S. M. Haas and J. H. Shapiro. Capacity of the multiple-input, multiple-output Poisson channel. *Kansas Workshop on Stochastic Theory & Control*, B. Pasik-Duncan, Ed., Lawrence, KS, October 2001.

[12] K. Chakraborty and P. Narayan. The Poisson fading channel. *IEEE Trans. Information Theory*, 53(7):2349–2364, July 2007.

[13] K. Chakraborty, S. Dey and M. Franceschetti. Outage Capacity of MIMO Poisson Fading Channels. *IEEE Trans. Information Theory*, 54(11):4887-4907, November 2008.

[14] K. Chakraborty. Capacity of the MIMO optical fading channel. *Int. Symp. Information Theory*, 530–534, Adelaide, Australia, September 2005.

[15] S. M. Haas and J. H. Shapiro. Capacity of wireless optical communications. *IEEE J. on Selected Area in Communication*, 21(8):1346–1357, October 2003.

# Biographies

**Razgar Rahimi** was born in Saghez, Kurdistan, Iran in 1981. He received the B.Sc. degree in electrical engineering from Iran University of Science and Technology, Tehran, Iran in 2005, and the M.Sc. degree in electrical engineering from Shahed University, Tehran, Iran in 2010. His recent research interests include cognitive radio networks and cooperative communication systems.

**Gholamreza Dadashzadeh** was born in Urmia, Iran, in 1964. He received the B.Sc. degree in communication engineering from Shiraz University, Shiraz, Iran, in 1992 and M.Sc. and Ph.D. degrees in communication engineering from Tarbiat Modarres University, Tehran, Iran, in 1996 and 2002, respectively. From 1998 to 2003, he has worked as the Head Researcher of the Smart Antenna for Mobile Communication Systems and the wireless local-area network 802.11 project with the radio communications group of Iran Telecommunication Research Center (ITRC). From 2004 to 2008, he was the Dean of the Communications Technology Institute, ITRC. He is currently an Assistant Professor with the Electrical and Electronic Engineering Department, Shahed University, Tehran. He has published more than 80 papers in referred journals and international conferences in the area of antenna design and smart antennas. Dr. Dadashzadeh is a member of the Institute of Electronics, Information, and Communication Engineers of Japan and the Iranian Association of Electrical and Electronics Engineers. He received the first degree of national researcher in 2007 from Iran's Ministry of Information and Communications Technology.

N. Werghi, N. Medimegh and S. Gazzah

# Watermarking of 3D Triangular Mesh Models Using Ordered Ring Facets

**Abstract:** In this paper we propose fragile and robust methods for watermarking binary data in triangular mesh models. Contrary to digital images and audio which benefit from the intrinsically ordered structure of the matrix and the array, 3D triangular mesh model lacks this capital property even though it can be encoded in an array date structure. Such a lack often complicates the different aspects of data embedding, like model traversal, data insertion, and synchronization. We address this problem with a mesh data representation which encodes the mesh data into a novel ordered structure, dubbed, the Ordered Rings Facets (ORF). This structure is composed of concentric rings in which the triangles are arranged in a circular fashion. This representation exhibits several interesting features that include a systematic traversing of the whole mesh model, simple mechanisms for avoiding the causality problem, and an efficient computation of the embedding distortion. Our fragile method can be also adapted to different scenarios of data embedding, which includes stenography and fragile watermarking. To maximize robustness of our robust method, we embed the signatures in stable regions where we have the minimum variance of spherical coordinates $\varphi$ detected using ORF. Experimental results show that this new approach is robust against numerous attacks.

**Keywords:** 3D triangular mesh watermarking, Robust watermarking, Fragile watermarking, Ordered Ring Facets.

# 1 Introduction

Data embedding refers to the process of inserting data into digital contents such as images, videos and 3D models. This process is referred to by different terms depending on the context, the scope and the aim of the data insertion. For example, for copyright protection, we refer to it by digital watermarking. Here the embedded data, called also the watermark, is meant to prove the ownership of the digital content and prevent its illegal use. Moreover, it is designed to resist and survive against

**N. Werghi:** Department of Electrical and Computer Engineering, Khalifa University, Sharjah UAE, email: Naoufel.Werghi@kustar.ac.ae.

**N. Medimegh and S. Gazzah:** Research unit on Advanced Systems in Electrical Engineering(SAGE), National Engineering School of Sousse University of Sousse, Tunisia, emails: sami_gazzah@yahoo.fr; Medimegh_Nassima@yahoo.fr

De Gruyter Oldenbourg, ASSD – Advances in Systems, Signals and Devices, Volume 4, 2017, pp. 75–91.
DOI 10.1515/9783110448399-005

malicious alterations (attacks). In such scenarios, the watermarking is labelled by the term robust. The term fragile watermarking is employed when the goal is to protect the integrity of the digital content from an unauthorized processing and to detect an eventual local or global manipulation. In fragile watermarking data is embedded globally (e.g. across the whole model surface), whereas in robust watermarking data is often locally embedded. Data embedding falls under the stenography scope when the digital content is used as a medium for carrying hidden information.

Within each of these applications, data embedding can be applied in the spatial or the spectral domain. In the former, the data is embedded by altering locally or globally the geometry or the topology of the model surface. Whereas the latter involves the modification of a certain components of a spectral transform coefficients. Data embedding can also be qualified to be blind or non-blind, depending on whether or not the original digital content is required to extract the embedded data.

With the recent advances of 3D shape acquisition and the CAD technology and the rapid evolution of network bandwidths, data embedding witnessed a new trend towards the use of 3D triangular mesh models. This fueled the need for new approaches and paradigms that can address the problematic aspects of embedding 3D models, and 3D triangular mesh models in particular.

A triangular mesh is a group of connected triangular facets that encode a given surface in terms of geometry and connectivity. The geometry defines the location of the triangular facets' vertices in the Euclidean space. The connectivity defines the sets of vertices that are connected to form the triangles or the facets of the mesh. Each triangular facet is referenced by an index value that points to the three vertices bounding the triangle. As has been mentioned by Wang *et al.* [34], there is no simple robust intrinsic ordering the mesh elements, e.g. facets and the vertices, which often constitute the carrier of the embedded data. Some intuitive orders, such as the order of the vertices and facets in the mesh file, and the order of vertices obtained by ranking their projections on an axis of the objective Cartesian coordinate system, are easy to be altered.

To address this lack of order and its consequent issues in data embedding, we propose to use a novel paradigm, dubbed the Ordered Ring Facets (ORF). The paradigm has been firstly introduced in [1, 2] in the context of 3D facial surface analysis. In this paper we showcase the adaptability of this paradigm to data embedding for 3D mesh models and its distinguished features.

The rest of the paper will be organized as follows: Section 2 reviews part of the literature work. Section 3 describes the ORF concept and the related algorithms. Section 4 exposes the different aspects of our watermarking methods. Some experimental results are also presented and discussed. Finally we draw some concluding remarks in Section 5.

# 2 Watermarking methods

As we mentioned in the literature, watermarking methods can be segmented into robust methods and fragile methods. These can be also classified into different sub-categories depending on the nature of the model alterations.

## 2.1 Robust methods

These methods can be classified into spatial methods and spectral methods. The spatial methods act either on the geometry or the topology of the model. The elementary entity carrying the watermark data is referred by the watermark primitive. Many methods operate on the triangular facet as a watermark primitive. Benedens [3] proposed a technique based on the Extended Gaussian Image, which is a kind of orientations histogram employed here for embedding data. The EGI has been augmented by an imaginary component by Lee and Kwon [4, 5], and dubbed it, the CEGI. They claimed the same level of impercibility than Benedens's methods, in addition to its robustness against remeshing, simplification, cropping and noising.

In another method, Benedens [6] proposed algorithm, dubbed vertex flood, whereby the distance to the center of mass of reference triangles is changed to encode the watermark bits.

Kuo *et al.* [7] introduced the moment-preserving method , in which groups of neighboring facets are selected and classified using some specific geometric moments in order to hold the embedded bits. Other type of methods used particular special volumes for inserting data bits. Harte *et al.* [8] presented a blind watermarking scheme in which the embedded vertices are confined within ellipsoid or rectangular volumes.

Some works performed data embedding using spherical coordinates $(r, \theta, \varphi)$ as was firstly proposed by Zafeiriou *et al.* [9]. In this work, he restricted the displacements to set vertices having the coordinate $\theta$ within specific ranges in order to achieve robustness against mesh simplifications.

Topological methods proceed with data embedding by altering the connectivity and other topologic features of the mesh. Mao *et al.* [10] suggested subdividing triangles and inserting the data in the newly formed vertices. This method is robust to affine transformation. The triangle flood algorithm [6] of Benedens generates a unique path in the mesh, based on topological and geometric information, along which the data is inserted.

Basically, spectral methods embed the data in certain coefficients of harmonic or multi-scale transform. Inspired by the Laplacian matrix-based mesh compression of Karni and Gotsman [11], several methods used the Laplacian coefficients for data embedding in different variants. Ohbuchi *et al.* [12] applied this paradigm to mesh model. They applied eigenvalue decomposition of a Laplacian matrix derived from mesh connectivity. Cayre *et al.* [13] developed a blind method employing piece-wise Laplacian decomposition.

Wu and Kobbelt [14] proposed a robust and fast spectral watermarking scheme for large meshes using a new orthogonal basis functions based on radial basis function. In order to enhance further the robustness against attacks, Praun *et al.* [15] developed a method based on the principles of spread-spectrum watermarking, previously employed in images, sound, and video. Yan Liu *et al.* [16] proposed the use of the Fourier-Like Manifold Harmonics Transform (MHT).

Multi-scale methods used mostly the wavelet transform (WT). Here data bits are inserted in the WT coefficients. Kanai *et al.* [17] non-blind method was among the first attempts in this category. They used the multi-resolution decomposition of Eck and al [18].

Ucchedu *et al.* [19] presented a novel algorithm whereby the vertices at a given resolution-level and the related coefficients at the same level are used for bit insertion. Yin *et al.* [20] rather employed Burt-Adelson pyramid decomposition . Hoppe *et al.* [24] proposed a multi-resolution framework based on the edge-collapsing operator.

Other authors such as [21–23] found advantages (increasing the capacity and enhancing the robustness) in using the spherical variants of the multi-resolution analysis such as [23] with its spherical parameterization, [21, 22] which proposed the use of spherical wavelet transform and [25] whom proposed the concept of Oblate Spheroidal Harmonics.

## 2.2 Fragile methods

Fragile watermarking is used for checking the authenticity and the integrity of 3D mesh model. In this context the watermark is meant to be sensitive to the least amount of mesh modifications as well as to indicate the locations of such modification in the mesh. As for robust watermarking, methods in the category can be segmented into spatial and spectral methods.

Yeung and Yeo [26, 27] pioneered fragile watermarking of 3D models for verification purposes by extending a 2D image watermarking to 3D. They proposed the idea of moving the vertices to new positions so that each vertex would have the same value for two different and predefined hash functions. Attacks can then be revealed by the presence of vertices that do not comply with aforementioned condition. In this method the hash function requires a predefined order of the vertices within the 1-ring neighborhood, otherwise the scheme becomes vulnerable to the causality problem. Ohbuchi *et al.* [28] method imbeds the data on a facet quadruples (a facet and its three adjacents) across the whole mesh. The quadruple facets must satisfy similarity conditions, dubbed, Triangle Similarity Quadruple (TSQ), that is used to recall them when the embedded information is retrieved. Each quadruple stores four symbols composed of marker, subscript and two information data. These are embedded in the dimensionless features of the triangles (e.g. edge ratios), modifying the vertices' positions. To avoid the causality problem the facet quadruples should not be connected to each other.

Lin *et al.* [29] approached the causality problem by proposing a rearrangement of the pixels harmless to the embedded watermark and making the two hash functions depending only on the position of the current vertex. Chou *et al.* [30] proposed a watermarking mechanism in which one of the hash functions is dependent on the mean of the 1-ring vertex neighborhood. This mean is kept stable after watermarking by adjusting a vertex associated to the watermarked vertex.

High capacity steganographic methods, where the integrity of the hidden data is a requirement, can be also classified as fragile. In these methods too, vertices are altered to embed data bits. The larger the number of bits, the higher will be the capacity of the method. Cayre and Macq [31] proposed a two-stage blind method where, at first they select a candidate stripe of triangles, and then perform a bit embedding by projecting a triangle summit on the opposite edge segmented into two equal intervals. A facet is assigned the bit 0 or 1 depending on in which segment the projection occurred. The synchronization used some local (e.g. largest facet) or global (e.g. facet intersecting the largest principal axes) geometrical features. Bors [32] proposed a blind watermarking method that locally embeds a string of bits on a set of vertices selected and ordered based on a certain distortion visibility criterion. The vertices associated to 0 (respectively 1) are shifted outside (respectively inside) a bounding volume. He proposed two variants, in the first one, the bounding volume is an ellipsoid defined by the principal axes of the covariance matrix computed over the set 1-ring neighborhood, whereas in the second, bounded parallel planes are used. Here the vertex is moved along or opposite to the plane's normal depending on the bit value assigned to it.

A fragile method acting on the mesh connectivity, dubbed Triangle Strip Peeling Symbol Sequence, was also introduced by Ohbuchi *et al.* in [28]. The method consists in cutting out a stripe from the mesh except the attaching edge that marks the start of the stripe. The stripe is formed by repeatedly appending adjacent facets through a path encoded in the message data. The stripe can be shaped as a meaningful pattern that becomes visible when the mesh undergoes global connectivity alteration. However, in this method the watermark cannot spread over the whole mesh, and therefore it cannot be employed for integrity authentication.

In the frequency space, geometrical wavelet transform has been an attractive tool. Here the watermark is inserted by altering the wavelet-transform coefficients computed at each facet or by altering the facets at a given wavelet transform resolution to equate a predefined function. Cho *et al.* [33] followed the latter paradigm by embedding the watermark data in facets of the lower resolution of the wavelet transform. This method suffers, however, from the causality problem. The method of Wang *et al.* [34] alters rather the module and the orientation of the one-level WT coefficients to keep a same watermark symbol across the whole facets. This scheme has been also extended to multi-resolution levels in [35].

# 3 ORF representation

The ORF representation is a structure in which triangular facets are arranged and ordered in a sequence of concentric rings that emanates from a root or seed facet. This representation has been inspired from the observation of the arrangement of triangular facets lying on a closed contour of edges (Fig. 1.a). We classify these facets into two groups: 1) the *Fout* facets, comprising facets having an edge on the contour and pointing outside the area delimited by the contour, and 2) the *Fin* facets, comprising facets having a vertex on the contour and that point inside the area delimited by the contour. We also notice that the *Fgap* facets seem to fill the gaps between the *Fout* facets. Together, these two groups form a kind of ring structure. From this ring, we can derive a new group of *Fout* facets that are one-to-one adjacent with their *Fgap* facets. These new *Fout* facets will, in their turn, form the basis of the subsequent rings (Fig 1.b). By iterating this process, we obtain a group of concentric rings. These rings can be centered on a specific facet by setting the closed ring to the edges of that facet (Fig 1.c). Moreover, the sequence of facets across each rings can be ordered clock-wise or anti-clockwise (Fig 1.d).

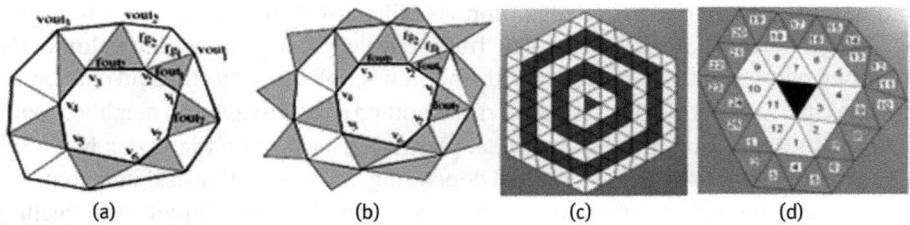

(a)        (b)        (c)        (d)

**Fig. 1.** a: *Fout* facets (dark) on the contour $E_7 : (v_1, v_2, ., v_7)$. The *Fgap* facets (clear) bridge the gap between pairs of consecutive *Fout* facets. b: Extraction of the new *Fout* facets. Notice that the new *Fout* facets are one-to-one adjacent to the *Fgap* facets. c: An example of a 5-ring ORF. d: facets in each ORF ring arranged.

The algorithm *ORF_Rings* has a computational complexity of $O(n)$ where $n$ is the number of facets in the rings. The function *GetRing* extracts the sequences of *Fgap* facets across the pairs of consecutive *Fout* facets, constructs the new ring and derives the *Fout* facets for the subsequent ring. The circular arrangement within one ring implicitly produces a spiral-wise ordering of the facets across the concentric rings.

The algorithm for constructing the ORF rings is as follows:

**Algorithm ORF_Rings**
Rings ← ConcentricRings(Fin_root, Fout_root)
    Rings ← [ ]; Fgap ←Fin_root ; Fout ←Fout_root
    For i = 1:NumberOfRings
        (Ring, NewFout, NewFgap) ← GetRing(Fout, Fgap)
        Append Ring to Rings
        Fout ← NewFout
        Fgap ← NewFgap
    End For
**End ORF_Rings**

# 4 Applications

This section exhibits the generality of the ORF framework by adapting it to 3D mesh watermarking. Section "Data embedding of 3D triangular mesh models using ordered ring facets" describes a novel fragile method for embedding binary data in triangular mesh models based on the ORF rings. Section "A Robust 3D Triangular Mesh Watermarking using ordered ring facets" exposes the watermarking process in 3D mesh model of a new robust method employing also the ORF structure.

## 4.1 Data embedding of 3D triangular mesh models using ordered ring facets

In this section we will describe the data embedding process in 3D mesh model employing the ORF structure and we will discuss its features and properties with respect to different data embedding scenarios. In the rest of the paper, we will refer to the embedded data by the term "payload" employed in stenography context.

### 4.1.1 Selection of the root facet

Usually, the choice of the first triangle is based on either local properties, e.g. the largest/smallest facet areas or global properties, e.g. the facet intersecting the major axis of the object principal axes. The method we propose exploits the ORF structure to set a secure starting facet that can be identified using a secret key, which we refer to by key 0. The key is a specific sequence of bits embedded in the 2nd and the 4th ring of the ORF. Therefore any facet can serve as a first key, provided it is marked by the secret key with which it can be identified. This secret key will be then part of the embedded data.

## 4.1.2 ORF bit embedding

The data embedding process uniformly spread the payload across the 3D mesh model by following the traversal path implicitly defined in the ORF rings. Indeed, the facets within the ORF rings are arranged in a kind of a spiral path that spans the whole model starting at the first facet. This ordering aspect allows data embedding in different ways depending on the nature of the application. For example, for model integrity verification, the payload can be inserted at equally spaced and ordered locations within this path. Integrity checking can then be easily and efficiently performed by parsing the path, retrieving and checking the payload at each specific location. For stenography applications, where the 3D mesh model is used to hide a secret data, the payload can also be spread across the whole path. Moreover, the ordered structure of the ORF allows different encrypting scenarios. In this paper, we will showcase a data embedding process within this latter scope.

Figure 2 depicts the different stages of the embedding process, illustrated over a sphere mesh model. First the ORF rings are extracted from the mesh model then they are sampled (sampling parameter $\tau$). The sampling aims to avoid having adjacent rings. Afterwards the rings' indexes are scrambled. This ring sampling and scrambling compose the first layer of encryption. Their related parameters (e.g $\tau$ and the scrambling code are stored in the encryption key labeled key 1. Afterwards a second layer of encryption is applied using a circular shifting of the facets in each ring followed by a sampling of order $\rho$. Each ring will have its own shifting value. The groups of shift values and the sampling parameter $\rho$ constitute the third encryption key (key 2), to be used, together with the key 0 and key 1, for retrieving the embedded data. Assuming a reasonably tessellated mesh, the set of facets that come out from stage 2 are uniformly spread over the mesh model surface. In the last stage the payload is embedded in this set at the cadence of one bit per facet. Here we employed the embedding technique of Cayre and Macq [31], but other bit embedding techniques can be used as well. In Cayre and Macq method, basically, one vertex is projected into its opposite edge, which is divided in two equal intervals, coded 1 and 0. If the projection falls in the interval that matches the embedded bit (for instance, it falls in the interval 1 and the bit to be embedded is 1), then the vertex is kept unchanged. Otherwise it is shifted collinearly with that edge, and within the facet's plane, to a new location for which the projection meets the good match.

The sampling performed in stages 1 and 2, aims to ensure isolation between the tampered, that carry the payload, so that they cannot share an edge or a vertex. The appropriate values of $\tau$ and $\rho$ depend on the regularity of the mesh. For a uniform mesh, $\tau = 2$ and $\rho = 3$ can guarantee compliance with the aforementioned condition.

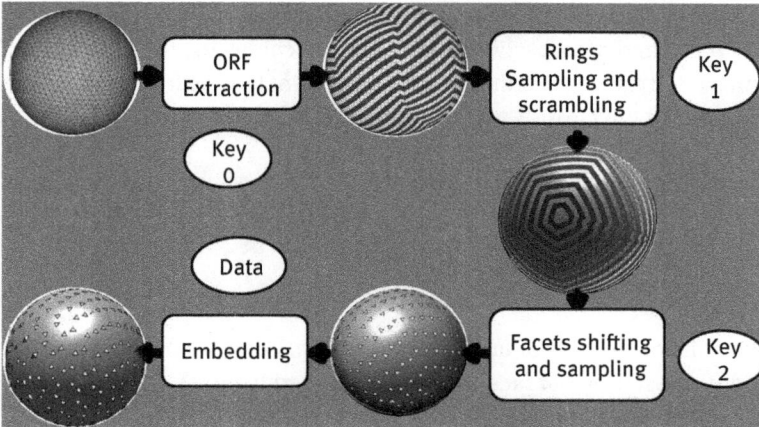

**Fig. 2.** Data embedding process

The uniform sphere mesh model depicted in Fig. 2 is an example of this case. When the mesh exhibits some irregularities, larger values might be needed.

### 4.1.3 Data extraction

The data extraction process is virtually similar to the embedding process. The starting facet should be firstly detected. Here the key 0 is used to inspect the four rings around the candidate facet. Afterwards, the rest of the ORF are extracted then browsed in the order and at the sampling rate encoded in Key 1. Then the facets in each ring are parsed according to the parameters encoded in key 2.

### 4.1.4 Causality problem

Our method offers control mechanism allowing data embedding free from the causality problem. Such a problem happens when an anterior tampered facet (e.g. already carried a bit) is to be affected by a posterior one (e.g a newly tampered). This might corrupt the content of the former facet. Usually this problem occurs between neighbouring facets. The size of the neighbourhood depends on the embedding technique and on how many facets surrounding the tampered one can be affected. This problem is addressed by inserting a flag data in or around the tampered facets so that these will not be embedded again when revisited during the mesh traversing. In our method, a proper setting of the sampling parameters $\tau$ and $\rho$ ensures sufficient spaces between the tampered facets preventing any mutual alteration, avoiding therefore the need for flagging and checking procedures. In the technique of Cayre and Macq [31], which we have used, only facets sharing the displaced vertex of the tampered facet are affected. Therefore, for a regular mesh, setting the sampling parameters $\tau$ and $\rho$ to 2 and 3, respectively, any triangles connected to the tampered facet, either by an edge or a vertex, cannot be the subsequent tampered one. However, depending on the state of the mesh, $\tau$ and $\rho$ might need to be increased.

### 4.1.5 Capacity

The total number of facets $N$ can be expressed by $N = \sum_{i=1}^{M} m_i$, where $M$ is the number of ORF rings and $m_i$ is the number of facets in the $i^{th}$ ring. Considering the sampling parameters $\tau$ and $\rho$, the number of facets can be expressed by $F = \sum_{k=1}^{M/\tau} m_{\tau k}/\rho$. The number of tampered facets can be evaluated in average to $N/(\tau \times \rho)$ bits. Assuming a uniform mesh in which the facets have close area and edge values, $\tau =$ and $\rho$ can be set to 2 and 3, respectively, the number of embedded facets can reach $N/6$.

### 4.1.6 Security

Accessing the payload requires addressing three challenges in our algorithm, namely, finding the root facet, finding the ring sampling rate $\tau$ and the correct combination of ring order, and finally finding the facet sampling rate $\rho$ and the facet shifting value at each ring. These three cascading stages of encryption give our scheme a high level of security. Exhaustive search is virtually out of reach, indeed the number of encryption combination in our scheme is evaluated to

$$N \times M! \times n_\tau \times \prod_{k=1}^{M/\tau} m_{\tau k} \times n_\rho \tag{1}$$

Where $N$ is the number of facets in the mesh, $M$ is the number of rings, $n_\tau$ and $n_\rho$ are the numbers of possible values of the sampling parameters $\tau$ and $\rho$, and $m_i$ is the number of facets in the $i^{th}$ ring. Such a number makes the extraction of the payload without the encryption code nearly impossible.

### 4.1.7 Mesh distortion

For mesh models with a single topology, Haussdorff distance has been adopted as an objective metric for evaluating the mesh distortion inferred by the data embedding. The computation of the Haussdorff distance is time demanding because of its quadratic complexity ($O(n^2)$). In our method, the computation of the mesh distortion is brought down to a linear complexity $O(n)$. In effect, the linear traversal path embedded in the ORF structure allows a one-to-one mapping between the original model and the embedded model facets. Based on this mapping, we can simply express the mesh distortion with the following formula

$$\sum_{i=1}^{N} \frac{\sum_{j=1}^{3} \min_{1 \le k \le 3} (u_i^j - v_i^k)}{3\bar{u}_i} \tag{2}$$

where $u$ and $v$ form the pair of corresponding facet edges in the original and embedded model respectively, $\bar{u}$ is the mean of facet edge in the original model, and $N$ is the number of facets.

### 4.1.8 Integrity checking

As the embedding is uniformly spread over the whole model, our method can be used for checking the integrity of the mesh model by comparing the extracted payload (formatted as a string of bits), and matching it with its reference counterpart. A mismatch indicates that the model has been altered. The indexes of the mismatches in the string of bits can be easily mapped to the facets' indexes, and thus used to trace the locations of the attacked areas. Moreover, the global mesh alteration can be evaluated using the following simple formula

$$\sum_{i=1}^{M} \delta_i, \ \delta_i = \begin{cases} 0 & \text{if } \Omega_i = Y_i \\ 1 & \text{otherwise} \end{cases} \tag{3}$$

where $\Omega$ and $Y$ are the original and the retrieved strings and $M$ is their length.

### 4.1.9 Robustness

Being dedicated to stenography applications, the method is robust to affine and similarity transforms, namely, translation, rotation, uniform scaling, and affine transforms. It is also resistant to vertices and facets reordering. It is not robust against geometrical or topological transformation such as cropping, simplification, and mesh resembling. These kind of attacks corrupt the message content.

### 4.1.10 Experiments

We applied on several mesh models collected from the 3D mesh watermarking benchmark [36]. The models are bunny, horse, rabbit, and venus. We used a PC-based

2.93 GHz, 8 GB RAM. The implementation was performed under Matlab. Table 1 depicts the list of the models, their corresponding sampling parameters, and the related performance measures. The embedding capacity represents the number of bits embedded in the model, which is also equal to the number of tampered triangles, since the embedding technique inserts one bit per triangle. The distortion is computed using equation 2. The different models show a capacity in the range of 7%–14%. However this could be further increased by inspecting further each ORF ring to locate other free facets. The distortion shows quite low values. It seems inversely proportional with the sampling parameters. This is expected as the number of tampered facets increases as the sampling decreases. The processing time seems to increase at a fair rate because of the linear complexity of our method. This aspect is confirmed in another experiment, performed with the horse model, in which we computed the embedding time for increasing sizes of the payload (Fig. 4). The growth rate of the processing time confirms clearly the linear complexity property of our method.

**Fig. 3.** ORF rings and the tampered facets displayed on the models.

**Tab. 1.** Results obtained for the tested models.

| Model | facets | vertices | $\tau$ | $\rho$ | capacity | distortion | time |
|---|---|---|---|---|---|---|---|
| bunny | 69666 | 34835 | 6 | 8 | 3201 | 0.90 | 6.85 |
| rabbit | 141312 | 70658 | 4 | 7 | 5121 | 1.65 | 8.89 |
| venus | 201514 | 100759 | 3 | 9 | 9016 | 0.56 | 13.13 |
| horse | 225080 | 112642 | 3 | 5 | 15845 | 3.35 | 13.48 |

## 4.2 A robust 3D triangular mesh watermarking using ordered ring facets

The proposed method embeds watermark into 3D triangular mesh model by modifying the component spherical $\varphi$ of some vertex according to assigned watermark bit and distribution of $\varphi$. The watermark embedding process is illustrated in Fig. 5.

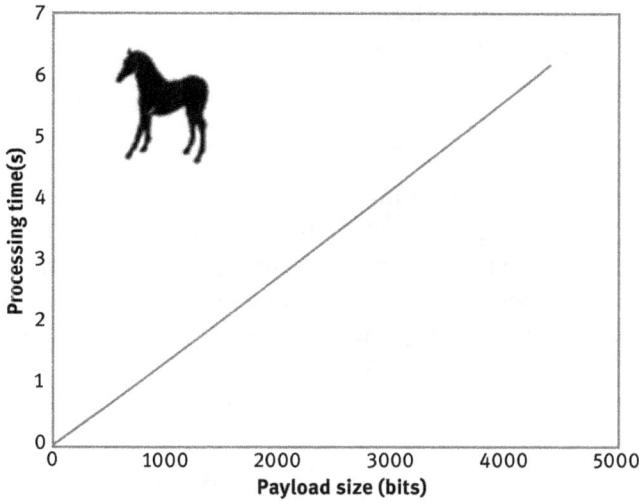

**Fig. 4.** Processing time evolution for embedding 4408 bits in the horse model.

### 4.2.1 Watermark embedding process

The general algorithm of embedding decomposes into several steps. We embed the watermark by altering vertex position. The signature can be destroyed before extraction by the affine transformation (rotation, translation or scaling) which displaces the vertex from their original coordinate. To address this problem we normalized the mesh by translating the mesh from its center to the center of gravity. Then, we align the principal components of the mesh with the standard axes given by principal component analysis PCA. Second, Cartesian coordinates of a vertex $v_i=(x_i, y_i, z_i)$ (for $1 \leq i \leq N$ where N is the number of the vertex) are converted into spherical coordinates $(r_i, \theta_i, \varphi_i)$. The proposed method uses only the radial distance $\varphi$ for watermarking. Note that we have verified that $\varphi$ is the most invariant to attacks. Third, we select a region where we have the minimum variance of spherical coordinate's $\varphi$. From this region, we choose the embedded vertex uniformly spread over the whole selected region using the ordered ring facets. As facets of each ORF ring are arranged in order, we have followed a defined path for insertion. This aspect facilitates also the recovery of signature embedded at spaced locations. For each facet, we generate a set of 10 rings. Then, we select the ring 1, 3, 5, 7 and 9 to be watermarked, in order to avoid having adjacent rings and facets. The ORF ring is formed by two groups of facets: *Fout* facets and *Fgap* facets. As the sequence of facets across each ring is ordered, we choose to embed a signature into *Fout* facets. This type of facets has an edge on the contour and a vertex $V$ pointing outside the area delimited by the contour. Only several selected $V$ vertices will be affected. The next step of the proposed watermark embedding is to insert signature (one bit per facet). The basic idea is to modify the component spherical $\varphi$ of this vertex function of the bit to be inserted and an interval $I$ defined as follows

$$I = [\mu - \sigma, \mu + \sigma] \tag{4}$$

Where $\mu$ is the mean and $\sigma^2$ is the variance of the components $\varphi$ of the vertices of each ring selected excluded the watermarking vertices. To embed a watermark bit of 1, vertex component $\varphi$ is transformed in order to be in the interval $I$. alternatively, to embed 0, vertex component $\varphi$ is transformed in order to move outside $I$. Finally, the watermark embedding process is completed by converting the spherical coordinates to Cartesian coordinates and by mesh renormalization.

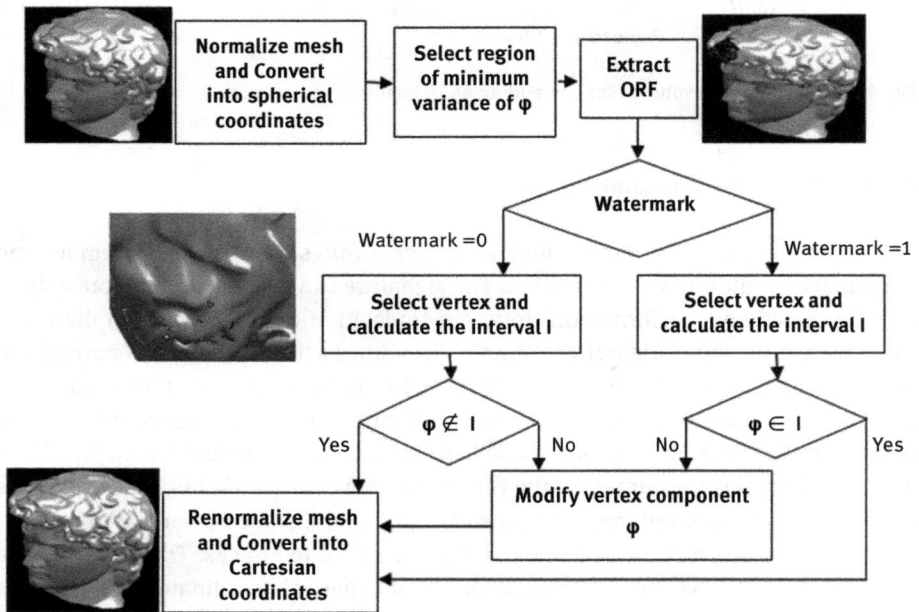

Fig. 5. Block diagrams of the watermark embedding for the proposed watermarking method.

## 4.2.2 Watermark extraction process

The algorithm of detection is similar to the watermark embedding process. The watermarked mesh model is first normalized and converted to spherical coordinates. After finding the region of minimum variance of $\varphi$, we exploit the ORF rings to detect the watermarked vertex. Then, we calculate the interval $I$. The watermark hidden W' is extracted finally by means of:

$$W' = \begin{cases} 1 & \text{if } \varphi \in I \\ 0 & \text{if } \varphi \notin I \end{cases} \tag{5}$$

## 4.2.3 Experimental results and evaluations

To evaluate the proposed approach, it is applied on a different mesh models used usually in 3D watermarking application. Only four examples are presented here: Bunny (12942 vertices, 25736 facets), Horse (7377 vertices, 14750 facets), Elephant (12471 vertices, 24950 facets), David_head(12865 vertices, 25640 facets). Naturally, the watermarking process should verify a good visual imperceptibility of the watermark. To measure the quality distortion, we used Metro [37]. This method consists to calculate the Hausdorff Distance $HD$ between the original mesh model and watermarked one. To measure the $HD$ the following formula is used:

$$HD = max\{h(M_1), h(M_2)\} \tag{6}$$

Where $M_1 = (V, V')$ and $M_2 = (V', V)$, ($V$ and $V'$ represent respectively the original mesh and watermarked mesh).
$h(M_1) = max\{min(d(a, V'))\}$, $a$ in $V$,
$h(M_2) = max\{min(d(b, V))\}$, $b$ in $V'$.

The robustness is measured by calculating the correlation between the original and the extracted signature.

$$corr = \frac{\sum_{n=0}^{N-1}(w_n - \bar{w})(w'_n - \bar{w}')}{\sqrt{\sum_{n=0}^{N-1}(w_n - \bar{w})^2 \times \sum_{n=0}^{N-1}(w'_n - \bar{w}')^2}} \tag{7}$$

Where $\bar{w}$ is the average of the signature and corr in $[-1, 1]$.

In the simulation, we embedded between 15 and 20 bits of watermark depending on the number of selected facet. The performance in term of $HD$ and correlation are listed in Tab. 2 when no attack. Here, the number of initial facet and the value of variance of the selected region before and after watermarking are also listed. It shows that we obtained the same region. So we can conclude about the stability of the selected region. The low result of $HD$ proves the good invisibility.

To evaluate the robustness, many attacks are tested. First, we have tested the watermark against the affine transformations (rotation, translation or scaling). We succeed to recover the signature after these three attacks. This is obtained through to normalization of the mesh. The next step, we evaluate the resistance to smoothing with different number of iteration. The watermark resists, if the parameter of smoothing is lower than 30 iterations. After that, we studied the robustness against noise attack by adding binary random noise to each vertex coordinates in the watermarked model. For the models tested, we can detect the signature after adding noise with different percentages. In the last step, we tested our method against simplification and subdivision attacks. However, we have a difficulty to detect the signature due to the adding or the simplification of the number of facet.

**Fig. 6.** Original mesh models followed by watermarked mesh models.

**Tab. 2.** Evaluation of watermarked meshes when no attack.

| Models | Corr | HD | $Var_\varphi B$ | $Var_{\varphi A}$ | Initial facet $B$ | Initial facet $A$ |
|---|---|---|---|---|---|---|
| Bunny | 1 | 0.0023 | 0.0023 | 0.0023 | 21441 | 21441 |
| David | 1 | 0.0160 | 0.0076 | 0.0076 | 21261 | 21261 |
| Horse | 1 | 0.0136 | 0.0008 | 0.0008 | | |
| | | | | | 1561 | 1561 |
| Elephant | 1 | 0.0078 | 0.0024 | 0.0024 | 15845 | 10241 |

# 5 Conclusion

In the paper, we have showcased new paradigms for embedding data in 3D triangular mesh model, based on Ordered Rings Facets. This representation is characterized by the innovative aspect of intrinsically embedding a structured and ordered arrangement of the triangular facets of the mesh surface. Our methods benefit from the advantages of this structure. Compared to other embedding methods, our methods are distinguished by an automatic traversal of the mesh model, an immunity mechanism against causality problem.

We applied this embedding paradigm to stenography and robust watermarking scenarios. The stenography method presents a linear complexity of the embedding process and a high encrypting power. It is also robust against translation, rotation, scaling, vertex and facets reordering. In addition to stenography, our paradigm can be easily adapted to fragile watermarking and authentication. The robustness of our robust watermarking method was verified after application of various types of attacks such as affine transformations (rotation, scaling and translation), smoothing and random noise added to vertex coordinates.

For the moment our watermarking scheme is not robust against sever topologic attacks. This is a point which we will address in a future work. Also, the embedding capacity is one of the aspects that deserves some improvements. At the current state, our paradigm ensures an embedding automatically free of the causality problem, but seemingly at the expense of the embedding capacity. We plan to further investigate this issue so that we can reach a better compromise. One technique which we are currently studying is replacing the fixed-rate sampling at the ring level, with a kind of adaptative sampling whereby the sampling step is adjusted according to the facet's neighbourhood in the ring. Finally our fragile approach is not qualified to deal with the genius 2 models, exhibiting holes and gaps.

# Bibliography

[1]  N. Werghi, M. Rahayem, J. Kjellander. An ordered topological representation of 3D triangular mesh facial surface: Concept and applications. EURASIP Journal on Advances in Signal Processing, 144, 2012.

[2]  N. Werghi, M. K. Naqbi. A novel surface pattern for 3D facial surface encoding and alignment. Proc. IEEE SMC, 902–908, 2011.

[3]  O. Benedens. Geometry-Based Watermarking of 3D Models. IEEE Computer Graphics and Applications, 46–55, January 1999.

[4]  J. W. Lee, S. H. Lee, K. R. Kwon, K. I. Lee. Complex EGI Based 3D Mesh Watermarking. IEICE Transactions on Fundamentals of Electronics, Communications and Computer Sciences 88(6): 1512–1519, 2005.

[5]  S.H. Lee, K.R. Kwon. A Watermarking for 3D Mesh Using the Patch CEGIs. Digital Signal Processing, 17(2):396–413, March 2007.

[6]  O. Benedens. Two High Capacity Methods for Embedding Public Watermarks into 3D Polygonal Models. In Proc. Of the Multimedia and Security Workshop at ACM Multimedia , Orlando, FL, 95–96, 1999.

[7]  C.T. Kuo, D.C. Wu and S.C. Cheng. 3D Triangular Mesh Watermarking for Copyright Protection Using Moment-Preserving. 2009 Fifth Int Conf on Intelligent Information Hiding and Multimedia Signal Processing, 136–139, 2009.

[8]  T. Harte and A. Bors. Watermarking 3d models. In Proc. of Int Conf on Image Processing (ICIP), Rochester, NY, USA, 3:661–664, Sep 2002.

[9]  S. Zafeiriou, A. Tefas, and I. Pitas. Blind Robust Watermarking Schemes for Copyright Protection of 3D Mesh Objects. IEEE Transactions on Visualization and Computer Graphics,11:596–607, 2005.

[10] X. Mao, M. Shiba, and A. Imamiya. Watermarking 3D Geometric Models through Triangle Subdivision. In P. W.Wong and E. J. Delp, editors, Proc. SPIE Security and Watermarking of Multimedia Contents III, 253–260, 2001.

[11] Z. Karni and C. Gotsman. Spectral compression of mesh geometry. Proc. SIG-GRAPH, 279–286, 2000.

[12] R. Ohbuchi, A. Mukaiyama, S. Takahashi. A Frequency-Domain Approach to Watermarking 3D Shapes. Computer Graphics Forum 21(3): 373–382, Sep 2002.

[13] F. Cayre, P.R. Alface, F. Schmitt, B. Macq, H. Maitre. Application of Spectral Decomposition to Compression and Watermarking of 3D Triangle Mesh Geometry. Signal Processing 18(4):309–319, Apr 2003.

[14] J. Wu, L. Kobbelt. Efficient Spectral Watermarking of Large Meshes with Orthogonal Basis Functions. Visual Computer 21: 848–857, 2005.

[15] E. Praun, H. Hoppe and A. Finkelstein. Robust Mesh Watermarking. ACM SIGGRAPH'99 Conference Proceedings, 1999.

[16] Y. Liu, B. Prabhakaran, and X. Guo. A Robust Spectral Approach for Blind Watermarking of Manifold Surfaces. The 10th ACM Workshop on Multimedia and Security, 43–52, 2008.

[17] S. Kanai, H. Date, T. Kishinami. Digital Watermarking for 3D Polygons Using Multiresolution Wavelet Decomposition. In Proc of the International Workshop on Geometric Modeling, 296–307, 1998.

[18] M. Eck, T. D. DeRose, T. Duchamp, H. Hoppe, M. Lounsbery, and W. Stuetzle. Multiresolution Analysis of Arbitrary Meshes. in Proc. of the ACM SIGGRAPH Conference on Computer Graphics'95, 173–180, 1995.

[19] F. Uccheddu, M. Corsini, and M. Barni. Wavelet based Blind Watermarking of 3D Models. In MM & Sec Proceedings of the 2004 workshop on Multimedia and security, New York, NY, USA, 143–154, 2004.

[20] K. Yin, Z. Pan, J. Shi, D. Zhang. Robust Mesh Watermarking Based on Multiresolution Processing. Computers and Graphics 25:409–420, 2001.

[21] J. Q. Jin, M. Y. Dai, H. J. Bao, Q. S. Peng, Watermarking on 3D Mesh Based on Spherical Wavelet Transform, J.Zhejiang Univ, 5:251–258, 2004.

[22] J.J. Qiu, D.M. Ya, B.H. Jun, and P.Q. Sheng,Watermarking on 3D mesh based on spherical wavelet transform, J. Zhejiang Univ. 5:251–258, 2004.

[23] L. Li, D. Zhang, Z. Pan, J. Shi, K. Zhou, K. Ye, Watermarking 3D Mesh by Spherical Parameterization, Computers and Graphics 28:981–989, 2004.

[24] H. Hoppe. Progressive Mesh. in Proc. of the ACM SIGGRAPH Conference on Computer Graphics'96, 99–108, 1996.

[25] J.M. Konstantinides, A. Mademlis, and P. Daras. Blind Robust 3D Mesh Watermarking Based on Oblate Spheroidal Harmonics. IEEE Transactions on Multimedia, 23–38, 2009.

[26] M.M. Yeung, B.L. Yeo. Fragile watermarking of three dimensional objects. Proc.IEEE Int. Conf. Image Processing, ICIP98, 2:442–446, 1998.

[27] B. Yeo and M. M. Yeung. Watermarking 3D objects for verification. IEEE Computer Graphics and Applications, 19(1):36–45, Jan.-Feb. 1999.

[28] R. Ohbuchi, H. Masuda, and M. Aono. Data Embedding Algorithms for geometrical and non-geometrical targets in three-dimensional polygonal models. Computer Communications archive 21(15):1344–1354, 1998.

[29] H.S. Lin, H. M. Liao, C. Lu, and J. Lin. Fragile watermarking for authenticating 3-D polygonal meshes. IEEE Transactions on Multimedia, 7:997–1006, 2005.

[30] C.M. Chou and D.C. Tseng. A public fragile watermarking scheme for 3D model authentication. Computer-Aided Design, 38:1154–1165, 2006.

[31] F. Cayre and B. Macq. Data hiding on 3-D triangle meshes. IEEE Transactions on Signal Processing, 51(4):939–949, April, 2003.

[32] A.G. Bors. Watermarking mesh based representations of 3D objects using local moments. IEEE Transactions on Image Processing, 15:687–701, 2006.

[33] W.H. Cho, M.E. Lee, H. Lim, and S.Y. Park. Watermarking technique for authentication of 3D polygonal meshes. Proc. of the International Workshop on Digital Watermarking, 259–270, 2005.

[34]   K. Wang, G. Lavoue, F. Denis, and A. Baskurt. A fragile watermarking scheme for authentication of semi-regular meshes. in Proc. of the Eurographics Short Papers, 2008.

[35]   K. Wang, G. Lavoue, F. Denis, and A. Baskurt. Hierarchical blind watermarking of 3D triangular meshes. in Proc. of the IEEE International Conference on Multimedia and Expo, 1235–1238, 2007.

[36]   K. Wang and G. Lavoue, F. Denis, A. Baskurt, X. He. A benchmark for 3D mesh watermarking. Proc. of the IEEE International Conference on Shape Modeling and Applications, 231–235, 2010.

[37]   P.Cignoni, C. Rocchini, R. Scopigno. Metro: Measuring error on simplified surfaces. Computer Graphic Forum, 17(2):167–174, 1998.

# Biographies

**Naoufel Werghi** received PhD in Computer Vision from the University of Strasbourg. He has been a Research Fellow at the Division of Informatics in the University of Edinburgh, Lecturer at Department of Computer Sciences in the University of Glasgow. Currently, he is Associate Professor at the Electrical and Computer Engineering Department in Khalifa University, UAE. His main research area is image analysis and interpretation where he has been leading several funded projects in the areas of biometrics, medical imaging, geometrical reverse engineering, and intelligent systems. He published more than 70 journal and conference papers.

**Nassima Medimegh** received her research masters in Intelligent and communicating systems; option Signals and Communicating Systems in 2013 and diploma informatics engineering from national school of Sousse, Tunisia in 2011. Her research interests are in the general areas of image processing, 2D and 3D watermarking.

**Sami Gazzah** is an Assistant Professor in the Department of Telecommunication at the Higher Institute of Computer Sciences and Communication Techniques of Hammam Sousse from University of Sousse. He received the engineering degree and the Ph.D degree both in electrical Engineering from National Engineering School of Sfax/Tunisia. His research interests include intelligent Systems and computer vision.

M. A. Charrada and N. Essoukri Ben Amara

# Development of a Database with Ground Truth for Historical Documents Analysis

**Abstract:** Standard databases play essential roles for evaluating and comparing results obtained by different groups of researchers. In this paper, a database of images derived from the Tunisian heritage is introduced. First, we present the importance of digitization for the preservation and the protection of patrimonial collections. Then, we describe the development process of our database, intended for the analysis of historical document images, in collaboration with the National Archives of Tunisia. In the second place, we demonstrate the importance of the ground truth for image databases and the annotation impact of these images. Thus, we present our approach for images annotation which consists in producing manually and collectively keywords for each image and defining a new model based on an ontology which allows structuring of the knowledge extracted from images.

**Keywords:** Ancient documents, digitization, image database, ground truth, annotation, ontology.

## 1 Introduction

Since the appearance of the writing, humanity had the need to note and transmit their knowledge considering their limited storage capacities. Therefore, they have exploited over time a diversity of physical supports for the information conservation such as the stones and the wood. Nevertheless, the appearance of the paper has set off a scientific and cultural renaissance seeing that the transmission of knowledge from one period to another and from one civilization to another has become easier. However, the paper has been always a fragile support which is exposed to various forms of damage and deterioration that may affect its quality and its legibility, hence the appearance of the great need for the storage and the preservation of heritage documents under favorable conditions in order to keep them intact as long as possible. For this reason, several organizations, which have appeared throughout the world, have been responsible for the collection and the storage of patrimonial documents such as national archives, libraries and research centers. The main objective of these organizations has been to

**M. A. Charrada and N. Essoukri Ben Amara:** Advanced System in Electrical Engineering Research Unit (SAGE), National Engineering School of Sousse, University of Sousse, Sousse, Tunisia, emails: mohamed_aymen_charrada@yahoo.fr, najoua.benamara@eniso.rnu.tn.

De Gruyter Oldenbourg, ASSD – Advances in Systems, Signals and Devices, Volume 4, 2017, pp. 93–113.
DOI 10.1515/9783110448399-006

implement different policies which are necessary for the preservation and the protection of the historical data. Over the years, the masses of the preserved funds have become increasingly enormous which has caused many access difficulties for several safeguarded documents. Indeed, the mass of documents often obliges each reader to browse a huge amount of pages before finding documents that contain the information sought for.

Considering the various degradations that threaten archival documents and which are mainly due to the fragility of the information support, the poor storage conditions and the aging of documents, the archival organizations have significant challenges to raise: reducing the mass of preserved funds, making these funds available to the public in the best conditions and protecting archival documents against the probable threats as well as restoring their contents. The optimal solution has been to scan these documents so as to ensure the preservation of the patrimonial collections, the diffusion of document images using the newest technologies of communication and data access, and the possibility of implementing a digital process which ensures image correction and improves their quality in order to provide the most accurate digital version of the original document.

Thus, archival document digitization should not be limited to a simple scanning operation and must be extended to a processing chain which will ensure the digitization of historical documents and the development of digital libraries in order to reduce the physical space occupied by these documents, to improve the conditions of access to the patrimonial collections and to facilitate the exploration and the exploitation of these documents as well as to implement a strategy for the degraded document restoration.

In this article, we present our contribution in terms of design and annotation of historical document image database. In the next section, we give an overview on the digitization impact for the ancient document safeguard and protection. Then, we introduce in section 2 the retained approach for our database design. In section 3, we describe the importance of the ground truth for databases and the different annotation approaches present in the literature. Whereas, we present in section 4 the adopted approach for the annotation of our database images.

# 2 Database development

In this section, we will first present an overview on the patrimonial fund digitization and then we will describe the followed approach for designing our database.

## 2.1 Overview of patrimonial fund digitization

A new era of massive production and transmission of knowledge is marked by the invention of printers and scanners. These new technologies have challenged the supremacy of physical supports, especially the paper, as the only support of transmission and diffusion of information [1]. In this sense, the appearance of the digital supports of information has offered many advantages to libraries and archives. In fact, digitization ensures the ease of reproduction of documents without a significant loss of information, the ease of conservation, preservation and diffusion of ancient documents by reducing their consultation frequency and consequently reducing the causes of wear, and the improvement and the ease of access to heritage documents in their digital versions using a set of automated tools with a high capacity (such as search engines). Also, it reduces the physical space dedicated to the storage of patrimonial documents and replaces it with digital storage supports as it allows us to benefit possibly from the various computer editing functions ensuring a better possibility of exploitation of these documents [2]. Figure 1 illustrates different stages of the historical document digitization chain [2].

In order to take advantage of the document digitization benefits, several projects have been launched by libraries, archives and research laboratories throughout the world. These projects have often led to the creation of data set collections or databases which contain raw or proceeded digital images [3]. These collections are generally characterized by a significant size and diversified content and they are often exploited for scientific, literary or historical reasons. Table 1 shows an example of digitization projects that have been launched in recent years. These projects have established many image collections that are extracted from the human heritage which have allowed the researchers, throughout the world, to develop and validate several contributions in terms of historical document processing.

Nevertheless, a few number of digitization projects have been dedicated to the treatment of Arabic heritage and especially the Tunisian heritage. In fact, the digitization projects in the Arabic world have often suffered from a lack of organization, a stochastic design and an absence of required resources.

In this context, and in order to meet the scientific needs related to the research tasks conducted in our "Advanced System in Electrical Engineering" research unit (SAGE) and related to the treatment of the Tunisian heritage (preprocessing, segmentation, indexation, watermarking . . .), we have concluded that the ancient image database existing in our research unit does not satisfy these needs and we have to enrich this image database so that it will be rich, with ground truth and that it should include, as far as possible, the different types of available archival documents, covering the majority of problems (degradations and defects) involved in the historical documents, which can be used by our unit members to validate their research works and which can be published in order to be available for the scientific research purposes.

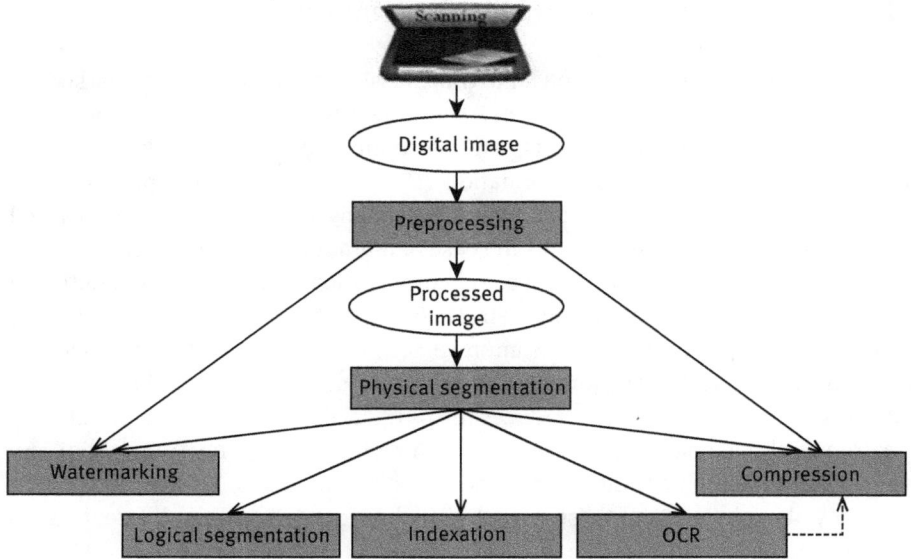

**Fig. 1.** Digitalization processing chain.

## 2.2 Database conception

In fact, our objective has been to design a database dedicated to the analysis of ancient document images. The development of such database requires resorting to organizations which provide access to this kind of documents.

**Tab. 1.** Exemples of international digitization projects.

| Project name | Organisation | Specifications |
|---|---|---|
| Google books [4] | Google company with partners | More than 20 million books from the 17th to the 19th century (2004–2012), Raw image diffusion |
| Gallica [5] | French National library with partners | 1 883 371 documents of various types (1997–2012), Raw image diffusion |
| Gramophone Virtuel [6] | Canadian Library and Archives | 1 500 078 documents of various types (1998–2012), Raw image diffusion |
| Madonne [7] | French National Research Agency | More than 100 000 manuscripts (2003–2006), Processed image diffusion |
| Navidomass [8] | L3i, LORIA, PSI, LI, CRIP5 and IRISA laboratories | More than 200 000 documents from the 16th century (2007–2009) Processed image diffusion |
| DiGidoc [9] | LaBRI, LI, LITIS, L3I Laboratories BNF, i2S and Arkhênum | More than 100 000 printed documents and manuscripts from the 16th to the 19th century (2011–2014), Processed image diffusion |

In this context, we have concluded a collaboration agreement with the National Archives of Tunisia (NAT) [10], which are responsible for the conservation, the protection and the use of the national archival heritage. This collaboration has allowed us firstly to have a direct access to the various documents in the archives as it provides us the opportunity to exploit the technical and the computer resources of NAT in order to perform digitization under the best conditions, as it has given us the opportunity to exchange information and experience with the NAT experts in terms of ancient-document processing and analysis.

At this level, the choice of documents to be digitized and the necessary criteria of digitization have presented a big challenge since they will carry obviously a possible influence on the rest of our work, and especially on the expected results. Therefore, we have decided to involve people who are interested in the ancient documents by providing a survey in order to benefit from the knowledge, the expertise and the real needs of these people. This survey, as well as the literature, has enabled us to formulate a set of specifications describing the digitization approach to follow. In fact, we have found, for example, that we must focus on Arabic manuscripts and periodicals because these two classes of documents provide gradually a scientific importance and are the subject of several recent research tasks. However, we could specify the necessary criteria for the digitization phase of the selected documents, allowing us to provide the more faithful images to the original documents. In this context and in order to conceive our database, we have decided to follow a four-step approach, used frequently for the digitization project planning, so as to plan our work, hence having satisfactory results. This approach is defined by the following steps.

### 2.2.1 Digitization project definition, referred objectives and target public description

During this step, we must determine the type of digitization project to carry out, choose the digitization parameters (size, format, resolution...) and set the project duration and the needed resources to achieve it. Our objective is to develop an image database of historical documents that will be used later by researchers for the development of patrimonial document treatment and restoration approaches [3, 11]. Based on the literature and on the available resources, the images of our database have been scanned:

- In grayscale mode (8 bits/pixel), which is recommended for most of monochrome, textual or photographic documents and which is sufficient for textual documents with a few colors; and in color mode (24 bits/pixel), which is needed for the color documents.
- At a resolution of 300 dpi which is acceptable for the most of ancient documents [12]. In fact, in the case of textual documents, this resolution is sufficient to reproduce the documents to their original size; whereas for the graphic documents or

to obtain enlargements (zoom), it is necessary to choose a higher resolution (600 dpi) [11].
–   With variable sizes considering that unlike recent documents which are standardized (e.g. A3 or A4), ancient documents have variable, and not standardized, dimensions [14].
–   Using the "TIFF" and "JPEG" formats. These formats are recommended for images in the majority of cases since they support the gray scale mode and the color mode as well as the lossless and lossy compression modes and the ability to store additional information [13].

Yet, this project will take place over a renewable period of three years in accordance with the progress of work.

### 2.2.2 Choice of documents to be digitized

This step consists in determining what types and which documents to be digitized and according to which criteria. The National Archive of Tunisia collects Tunisian historical documents, where the oldest ones date back to the seventeenth century. The funds are divided into the following categories:
–   Ottoman period (1574–1881);
–   Colonial period (1881–1956);
–   Independent Tunisia (since 1956);
–   Documentary funds.

Indeed, we have been able to collect documents belonging to the following classes:
–   Newspapers and periodicals in Arabic, French, Italian and Judeo-Arabic;
–   Manuscripts in Arabic and French;
–   Printed documents in Arabic, French and Italian which represent books, funds, monographs and administrative documents;
–   Arabic and French composite documents;
–   We also have found technical documents, forms, musical notes, and maps.

To fix the choice of the documents to be digitized, we conducted a study of the recent research works on the treatment of old documents as we exploited the survey already mentioned in order to define the needs and the suggestions of experts in terms of required document and information types.

### 2.2.3 Requirements and the available resource analysis

In fact, we need an image database covering the majority of problems related to old documents, having a consistent size and a diverse content. Nevertheless, the convention with the NAT has given us the opportunity to exploit the technical resources and the existing data in the NAT (especially in the NAT scanning room) in order to perform image digitization in the best conditions.

### 2.2.4 Document digitization

Indeed, the digitization of collected documents proceeds in the NAT local under the supervision of SID team members (signal, image and document team) of our research unit. Figure 2 shows an example of images taken from our database. After the development of our image database, we have had the need to classify and to annotate images of our database in order to provide a quick and fluid access to the treated documents and to reduce the time access to the existing images. Thus, an image database can be valuable and so be exploited by the researchers for the evaluation and the benchmarks of their approaches only if it is with ground truth or annotated. Therefore, we will present in the next section our contribution to the annotation of our database images.

# 3 Database with ground truth

In this section, we present first the different annotation techniques used in literature and then we describe the adopted approach for the annotation of our database images.

## 3.1 Image annotation

To facilitate the organization and the navigation in the large image collections and to benefit from advanced search techniques for images, it is essential to define the tools which make it possible to find the documents sought for through a set of criteria selected in minimum time. This is realizable by associating information (a set of descriptors or annotations) with the document images [18]. This annotation can be textual through a name, a place, a date, a keyword, etc. or geometric by referring to a position in the image, such as a box, a field, an area, etc. [14].

In fact, there exist several annotation techniques in the literature which are generally differentiated by the method used for the generation of image descriptors and which can be classified mainly in three categories.

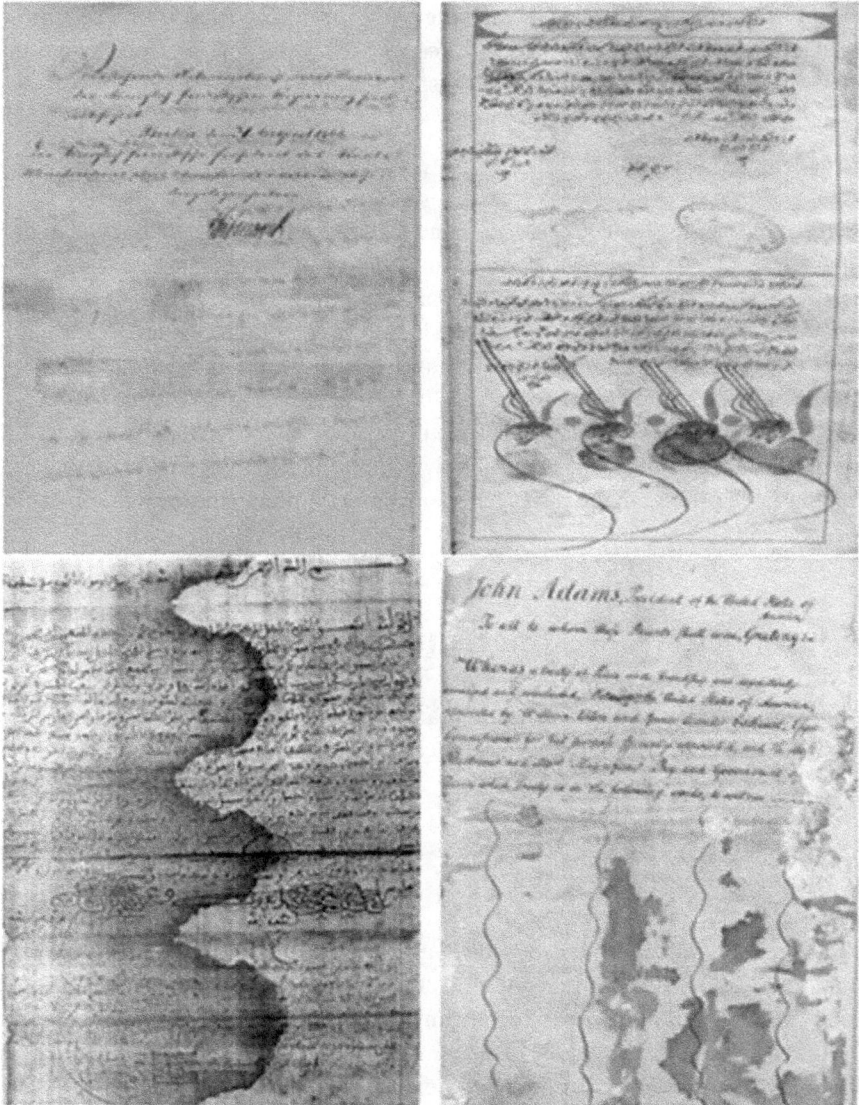

**Fig. 2.** Example of images taken from our database.

### 3.1.1 Manual annotation

This method consists in reading and decrypting the document by humans and performing the manual image description which facilitates the treatment of any kind of documents as well as obtaining the desired abstraction level. However, manual annotations are too subjective since they may vary from one person to another as they can become particularly tedious, too much long, with a significant risk of errors and especially with a particularly excessive cost according to the data quantity [15].

### 3.1.2 Automatic annotation

It consists in producing the annotations relating to the content of images in an automatic way using the document recognition techniques. This method is less expensive than the manual annotation but it requires the clarity and the sharpness of the document content for its proper functioning. However, this method does not work on old archival documents, because they are not sufficiently structured in the most of cases, and they generally suffer from a set of defects and degradations which may taint their contents [16, 17].

### 3.1.3 Collaborative annotation

This method allows combining the two previous types of annotation in order to profit from their advantages. It is generally suitable for old archival documents and consists first in locating manually interesting areas in the image and then automatically generating annotations from these areas [14].

For any selected annotation type, the ground truth generated for a database has to respect a set of requirements that will valorize the database. Indeed, the annotation vocabulary must be rich and representative to all database images, this representation must be valid visually and for the image content. Similarly, we must be careful about the required objectivity level of annotation; this level will represent the differentiating factor between one database and another. Table 2 shows an example of existing annotated databases.

**Tab. 2.** Example of existing annotated databases.

| Project name | Image number | Ground truth | Keywords/image |
|---|---|---|---|
| Graphic Obsession [19] | 500 000 | Manual | 10 to 50 keywords per image |
| DMOS [14] | 60 000 | Automatic (using DMOS method) | Average of 27 index per image |
| ARMARIUS [20] | 170 000 | Collective | 15 to 45 index per image |

In the next section, we present the retained approach for the annotation of our old document image database.

## 3.2 Description of our annotation approach

Generally, the selected type of the annotation depends on the required objective. In fact, our objective is to structure our image database as several collections of images according to a set of descriptors which make it possible to distinguish between images of different collections in order to facilitate the exploitation and the access to these images. In this sense, we have chosen to classify our database images based primarily on high-level features which represent our produced annotation knowledge. In fact, ancient documents often suffer from a set of degradations and defects that can be introduced at the stage of image capture such as inclinations, curvatures, lighting variation, blurred contours and parasitic points, as they may be due to the poor conditions of storage and conservation such as humidity tasks, burrs, and tears. Even the physical structure of ancient documents can introduce defects, which affect the quality and the legibility of these documents, such as the weak spacing between the various blocks and the variability of the page layout and styles in the document. These degradations represent an obstacle that makes the generation of automatic annotations with access to the document content a difficult task for many old documents. In addition, the automatic generation of annotations does not provide an acceptable level of abstraction and introduces the problem called "semantic gap" which represents the difference between the calculation of low-level descriptors (color, texture, shape) and the object identification in the image [21]. Therefore, we have chosen to apply the procedure of manual annotation since it is better suited to the nature of historical documents (especially for manuscripts), as it provides a higher level of abstraction. Then, to reduce the cost, the subjectivity level, the risk of errors and the length of manual annotations, we have had the idea of producing these annotations by several people (members of the SAGE research unit, visitors of the reading room of the NAT ...) so that each person can produce some annotations during their document consultation. Generally, the number of the produced annotations will naturally increase with the growing number of participants; accordingly the use of a selection and classification tool reduces the annotation mass and retains the most effective criteria and keywords.

So, our annotation procedure consists in extracting information from our database images (the annotation classes, descriptors ...) and restructuring the selected annotation data using an ontology in order to give meaning to the extracted information.

### 3.2.1 Annotation classes and keywords definition

As we mentioned previously, the annotation classes and keywords have been produced by several people. Thus, some annotation classes and keywords have been often quoted, and others, have been more subjective and rarely quoted. Therefore, we have gathered the produced data and we have removed those which are subjective in order to create the final list of annotation classes and the keywords by image. In this case, different views are well expressed in the annotations which makes the annotation more objective and the access to the images becomes more possible for a larger number of people. In what follows, we will present some produced annotation classes and keywords:

- *The historical era to which the document belongs:* Othoman, colonial and Tunisian independence periods;
- *The document type:* printed documents, manuscripts, periodical documents, maps, composite documents, musical notes, monographs . . .;
- *The size of the document image;*
- *The compression:* yes or no and the compression type;
- *The image resolution:* 200 dpi, 300 dpi, 600 dpi;
- *The image format:* JPEG, TIFF;
- *The language:* Arabic, French, Italian, Judeo-Arabic;
- *The document structure:* linear, hierarchical simple, complex . . .;
- *The types of degradations existing in the document:* humidity, acidity, parasitic points, bleed-through, inclinations, curvature, low contrast, luminosity variation, tears . . .;
- *The image colors:* black and white, grayscale, color;
- *The image content:* text, image, drawing, horizontal or vertical nets, array, tampon, frame, drop caps, mathematical formula . . .;
- *The content orientation:* vertical, horizontal, multi-oriented, inclined . . .;

It should be noted that the produced knowledge by the questioned public are high-level information and this is due to the nature of the questioned public which is composed of a set of students, archivists, historians and employees. In reality, it is difficult to meet appropriately to the expectations of different research areas. For example, the knowledge of historians makes it possible to describe images according to their expectations (historical value, type of the document, period, theme . . .), while the knowledge of image processing researchers allows the description of images according to criteria which are more appropriate to their discipline. So, we have retained the idea to model the obtained knowledge, presented in this case by the generated annotation classes and keywords, in order to reduce the problem of semantic gap. Such modeling is conducted using an ontology.

### 3.2.2 Ontology definition

The ontology is a structured set of concepts that give a meaning to information. Generally, it represents a data structuring based on description logics. From this structure, many rules are defined to constrain the coherence between all the data, and finally inference rules are used to deduce knowledge from the considered data. Thus, the ontology captures the relevant knowledge of a field, provides a common comprehension of the knowledge of this field, determines the recognized vocabulary of this field, gives an explicit definition of this vocabulary and the relationships between terms of the vocabulary, and all that in the form of formal models [22]. So, we have chosen to use this formalism in our work. The ontology is usually composed of [22]:

- *Individuals:* basic objects;
- *Classes:* sets, collections, or types of objects;
- *Attributes:* properties, features, characteristics or parameters that objects can own and share;
- *Relations:* the links which objects can have between them;
- *Events:* changes undergone by attributes or relationships.

Figure 3 illustrates the obtained class hierarchy used in our produced ontology.

Once the basic features and classes of images are identified, we assigned a role to each defined concept. The properties are used to ensure the coherence of the ontology, while the roles are used to extract knowledge. In the following, an example of roles and properties is shown.

- Document **HasAnEra**: Indicates the historical era to which the document belongs: "Document" is an individual, "Era" is property and "HasAnEra" is a role (relation).
- Document **HasALanguage**: Indicates the language(s) used in the document.
- Document **HasAStructure**: Indicates the structure of the document.
- Document **HasAContent**: Indicates the content of the document.
    - Content **HasAnOrientation**: Indicates the orientation of the document content.
- Document **HasAnImage**: Indicates the image of the document.
    - Image **HasAQuality**: Indicates the quality of the document image.
        - Quality **IsDegraded**: Indicates whether the image is degraded or not.

In fact, ontology is defined according to a specific formalism from which we can extract several requests for the access to our database images. In order to provide implementations of this formalism, we have chosen to use XML files to associate the generated descriptions with corresponding images. Figure 4 illustrates an example of an image with ground truth extracted from our database.

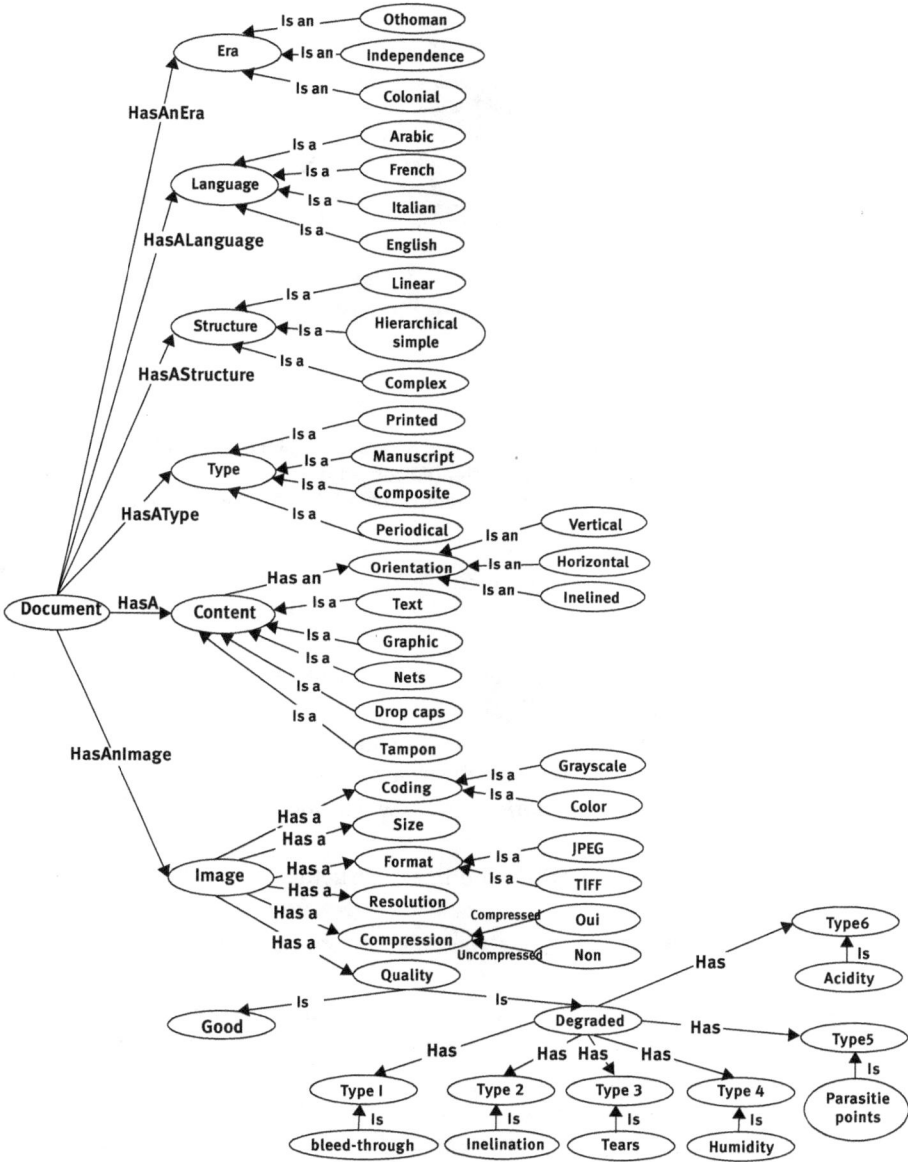

**Fig. 3.** Example of images taken from our database.

Currently, our database contains 10 000 images which are manually annotated using 970 keywords spread over about fifteen annotation classes with typically 15-30 keywords per image. These keywords are associated with each image using an

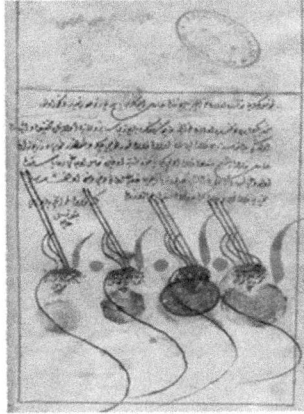

**Ground truth :** {Ottoman, 1851, manuscript, printed, 5504x4104, 300 dpi, JPEG, uncompressed, Quran, Arabic, simple structure, color, multi-oriented contednt, text, graphics, tampon, humidity, tears, low spacing between lines, fluctuating arrangement of the text, layout variation, parasitic points, luminosity variation}

**Fig. 4.** Example of ground truth taken from our database.

appropriate XML file allowing a better description and an easier representation of the various images. An example of such XML file is given in Fig. 5.

In fact, the used XML file is composed by six principal markup sections:

- *Description:* In this element, we have the transcription, the type and the language of the image as well as the historical era to which the document and its edition date belong.
- *Structure:* This element describes the image structure (layout) and its content orientation.
- *Content:* In this element, we quote the different physical entities present in the document image.
- *Degradation:* This element describes the various degradations present in the image. These degradations are classified into three categories described at the beginning of this section.
- *Specs:* In this element, we present the encoding of image, width, height and eventual addition effect. In the current version of our database, there are actually no added effects but we have planned to use this attribute for later versions of image rendering where effects could be present.
- *Generation:* This element presents the image resolution as well as the image colors, whether the image is compressed or not. For a compressed image, we specify the type of compression.

```
<?xml version="1.0" encoding="ISO-8859-1"?>
<img id = "1" name = "img1" >
  <description language = "arabe" transcription = "manuscrit"
type ="annexe passeport-turc " period = "ottomane"/>
  <structure>
    <oranisation> simple </oranisation>
    <orientation> horizontal </orientation>
  </structure>
  <content>
    <entity>
      <name> texte </name>
    </entity>
    <entity>
      <name> graphique </name>
    </entity>
  </content>
  <degradation>
    <capture>
      <name> variation de luminosité </name>
    </capture>
    <capture>
      <name> points parasites </name>
    </capture>
    <conservation>
      <name> humidité </name>
      <name> acidité </name>
    </conservation>
    <degphy>
      <name> faibles espacements entre les lignes </name>
      <name> disposition fluctuante du texte </name>
      <name> variation de la mise en page </name>
    </degphy>
  </degradation>
  <specs encoding = "jpg" width = "4104" height = "5504" />
  <generation date = "1851" resolution ="300 dpi" compression ="none" color = "yes" />
</img>
```

**Fig. 5.** Example of XML files including ground truth information about the image of Fig. 4.


# 4 Conclusion

In this paper, we have firstly presented the importance of digitization for the preservation and the protection of patrimonial collections and the reasons of the launching of massive digitalization projects by the archival organizations as well as the general approach followed during these projects. Thus, we have described the development process of our database, intended for the analysis of historical document images, in collaboration with the National Archives of Tunisia. Secondly, we have demonstrated the importance of the ground truth for image databases and the annotation impact of these images. Then, we have presented our approach for image annotation which consists in producing manually and collectively keywords for each image, provided by people having various statutes (students, researchers, historians, archivists...), which has allowed us to reduce the cost and the subjectivity of annotations, and we have tried to incorporate these information in the context of an ontology allowing a better description of the produced knowledge. In addition, we intend to continue the work in order to improve the used annotation syntax by low-level observations of images with an aim of reinforcing the coherence of image descriptors, provided by different people, according to the development of the database size and to enrich our ontology with an analysis and knowledge extraction module based on image processing defining a collaborative annotation approach that will consolidate manual and automatic annotations adapted to the nature of ancient documents in order to obtain a higher level of abstraction and to reduce the semantic gap of our annotation approach.

**Acknowledgment:** Special thanks to the Tunisian National Archives (NAT) for giving us the opportunity to access to its large document images database of Tunisian historical documents as well as for the helpfulness of its employees.

# Bibliography

[1]   F. Drira. *Contribution to the restoration of old document images*. PhD thesis, National Institute of Applied Sciences of Lyon, pages 16-38, mar 2007.

[2]   M. A. Charrada. *A segmentation approach development for old documents with complex structures*. PhD thesis, National Engineering School of Sousse, page 15, sep 2010.

[3]   M. A. Charrada, A. Kricha, and N. Essoukri Ben Amara. the treatment and the restoration of ancient documents. Arab Journal of Information, feb 2013.

[4]   Google. Google books, jan 2013. www.books.google.com/.

[5]   National Library of France. gallica, jan 2013. www.gallica.bnf.fr/.

[6]   National Library of canada. gramophone, jan 2013. wwww.collectionscanada.gc.ca /gramophone/.

[7]   L3I. Madonne, jan 2013. www.l3i.univ-larochelle.fr/MADONNE.

[8]   INRIA. Madonne, jan 2013. www.navidomass.univ-lr.fr/.

[9]   ANR. Armarius, jan 2013. www.agence-nationale-recherche.fr/projet-anr/.

[10]  ANT. Archives, jan 2013. www.archives.nat.tn/.

[11]  G. Bellemare, I. Cadrin, and M. Desroches. *The digitalization of the administrative documents-Methods and recommendations*. Archives branch, Library and Public archives of Quebec, jun 2010.

[12]  J. M. Ogier. Ancient document analysis: A set of new research problems. *6th French International Conference on the Writing and the Document (CIFED 2008)*, pages 151–166, oct 2008.

[13]  S. Ben Moussa. *Using fractal geometry for document analysis*. PhD thesis, National Engineering School of Sfax, pages 108-111, mar 2010.

[14]  B. Coüasnon and J. Camillerapp. *Access by the content to the handwritten documents of digitized archives*. Digital document, pages 61–84, jul 2003.

[15]  L. Hollink, G. Schreiber, and B. Wielinga. *Query expansion for image content search*. Pattern Recognition, pages 210–222, avr 2006.

[16] F. Le Bourgeois, H. Emptoz, E. Trinh, and J. Duong. Networking digital document images. *First International Conference on Document Analysis and Recognition (ICDAR 2001)*, pages 379–383, sep 2001.

[17] Y. Liu, D. Zhang, G. Lu, and W. Y. Ma. *A survey of content-based image retrieval with high-level semantics*. Pattern Recognition, pages 262–282, nov 2007.

[18] N. James and C. Hudelot. *Towards a semantized learning base for automatic annotation of large-scale images*. MAS laboratory, Central School Paris, June 2009.

[19] graphicobsession, jan 2013. www.graphicobsession.fr.

[20] R. Doumat, E. Egyed-Zsigmond, and J. M. Pinon. Model of collaborative numeric library-armarius. *6th International Conference on Information Research and Applications (CORIA 2008)*, pages 417–424, mar 2008.

[21] C. Millet. *Automatic images annotation: coherent annotation and automatic creation of a learning base*. PhD thesis, Paris 13 University, Paris-France, pages 16–22, Jan 2008.

[22] M. Coustaty. *Contribution à l analyse complexe de documents anciens Application aux lettrines*. PhD thesis, University, LaRochelle-France, pages 94–112, Oct 2011.

# Biographies

**Mohamed Aymen Charrada** received the computer engineering degree from the University of Sousse, Tunisia, in July 2008 and the master degree from the National Engineering School of Sousse, Tunisia, in September 2010. He is currently pursuing the Ph.D. degree in image processing at the University of Sfax.

**Najoua Essoukri Ben Amara** is a Professor with the department of industrial electronics of the National Engineering School of Sousse, Tunisia. She is currently serving as the director of the research unit "Advanced Systems in Electrical Engineering". She has broad research interests within the general areas of multimode Biometrics, pattern recognition, treatment of patrimonial documents.

T. Fei and D. Kraus

# Dempster-Shafer Evidence Theory Supported EM Approach for Sonar Image Segmentation

**Abstract:** In this paper, an expectation-maximization (EM) approach assisted by Dempster-Shafer evidence theory for image segmentation is presented. The images obtained by synthetic aperture sonar (SAS) systems are segmented into highlight, background and shadow regions for the extraction of geometrical features. Firstly, the proposed approach chooses the likelihood function given by Sanjay-Gopal *et al*. This likelihood function decouples the spatial correlation between pixels far away from each other. Secondly, the mostly implemented Gaussian mixture model is substituted by a generalized mixture model which adopts the Pearson system. As a consequence, the proposed approach is able to approximate the statistics of sonar images with more flexibility. Moreover, an intermediate step (I-step) is introduced between the E- and M-steps of the EM algorithm. The I-step adopts the Dempster-Shafer evidence theory based clustering technique to consider the spatial dependency among neighboring pixels. The states of neighbors are viewed as evidence to support the hypotheses about the state of the pixel of interest. Finally, the proposed approach is applied to SAS imagery to evaluate the performance.

**Keywords:** Image segmentation, Pearson system, Dempster-Shafer theory, Synthetic aperture sonar, Expectation-maximization algorithm.

# 1 Introduction

## 1.1 Problem statement

With the technological maturity of synthetic aperture sonar (SAS) systems, increasing attention has been attracted to the processing of SAS images for the sake of underwater mine countermeasures. The automatic target recognition (ATR) involved in underwater mine countermeasures contains mainly four steps as shown in Fig. 1: mine-like object (MLO) detection, image segmentation, object feature extraction and mine type classification. Many techniques [1–4] have been developed for object detection in the literature. The sonar images are inspected and the regions probably containing

**T. Fei, D. Kraus:** Institute of Wateracoustics, Sonar Engineering and Signal Theory (IWSS), Hochschule Bremen, Germany, emails: Dieter.Kraus@hs-bremen.de, fterryfei@gmail.com.
**T. Fei:** Hella KGaA Hueck & Co., Lippstadt, Germany. Email: Tai.Fei@hella.com.

De Gruyter Oldenbourg, ASSD – Advances in Systems, Signals and Devices, Volume 4, 2017, pp. 115–129.
DOI 10.1515/9783110448399-007

MLO are found. These regions are called regions of interest (ROI). The ROI are extracted and forwarded to the subsequent steps, i.e. the image segmentation and the feature extraction. Image segmentation is taken place in the second step to divide the images into highlight, shadow and background regions. Together with the ROI, the segmentation results are forwarded to the next step for feature extraction. Feature extraction and the associated feature selection techniques have been intensively studied in the literature, cf. [5–12]. Finally, those most relevant features are adopted by the last step of mine type classification. Classifiers (e.g. [13–18]) are trained and used to classify the types of those MLO.

According to the processing chain in Fig. 1, the results of image segmentation have a great influence on the feature extraction of MLOs. Hence, a reliable image segmentation algorithm is required. This paper has proposed a statistics based segmentation approach, which is assisted by the Dempster-Shafer evidence theory.

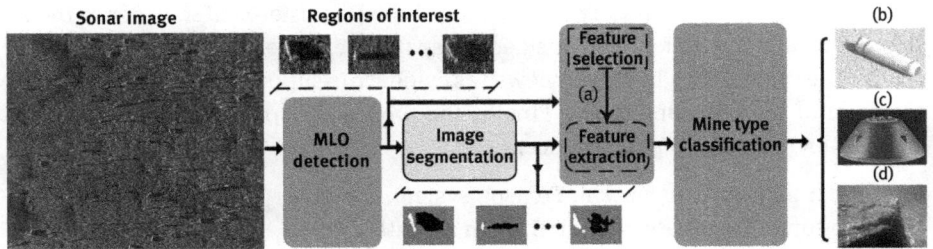

**Fig. 1.** An example of ATR system. The feature extraction step uses the results of both MLO detection and image segmentation as its inputs. (a) The results of feature selection instruct the feature extraction. The input of the system is sonar imagery and the output is the prediction on the types of the MLO: (b) a cylinder mine, (c) a truncated cone mine or (d) a stone.

## 1.2 Related work

In the last four decades, numerous works have been dedicated to the image segmentation. Simple techniques, such as thresholding [19–22] and $k$-means [23], have attained success for images with high signal-to-noise-ratio (SNR). The energy based active contour, e.g. [24, 25], is another popular approach image segmentation. However, according to our investigations, it is not optimal for the application in sonar images. Moreover, the statistics based approaches [26–28] have employed maximum *a posteriori* probability estimation to fulfill the task of image segmentation. The posterior probability function usually contains two parts to describe the conditional probability of the image pixel intensities given the class labels of pixels and the spatial correlation between the labels of neighboring pixels, respectively. A Markov random field (MRF) approach is mostly involved [29] in the posterior probability function to cope with the spatial dependency between pixel labels through the implementation of a Gibbs distribution. The setting of parameters adopted in Gibbs distributions for controlling the relationship between neighboring pixels is still open. Usually, they are set according to the experience gathered from specific applications. Mignotte *et al.* in [27] have used a least squares technique to estimate the parameters. This estimation requires the histogramming of neighborhood configurations, which is a time-consuming process. Besides, the conditional probability of image pixel intensities is typically modeled by Gaussian, gamma and Weibull distributions, which are often not adequate to approximate the statistics of the data obtained from real measurements.

The expectation-maximization (EM) algorithm [30] has been acting as a popular image segmentation approach for a long time, cf. [31, 32]. In order to consider the spatial correlation between neighboring pixels, Zhang *et al.* [31] substitute the pixel class probability provided by the M-step of previous iteration with an MRF based estimate. Later, Boccignone *et al.* [32] construct by inserting an anisotropic diffusion step [33] between each E- and M-step the so-called diffused expectation-maximization (DEM) scheme. With the assistance of the *a priori* knowledge that neighboring pixels are likely to be assigned with the same labels, neighboring pixels should have similar probabilities in the mixture distribution model. An anisotropic denoising filter is applied to probability levels so that the outliers with respect to their neighborhood are excluded, while the real edges of the image are still preserved. The application of such a denoising filter in DEM is not able to reliably exclude all of the noisy clusters in sonar images due to the fact that the variation of pixel intensities is high even for neighboring pixels. It is also possible to enlarge the object region because of the blur effect of denoising filters.

Most recently, the Dempster-Shafer evidence theory has been employed for image segmentation [34–36]. In [34–36] the segmentation of color images is considered, which can be divided into image components of R, G and B. These three image components are used as information sources. Hence, the authors in [34, 35] compose their belief structures based on the assumption of a Gaussian distribution. The mean and variance of the Gaussian distribution are estimated with the help of a simple thresholding technique [37] for each class. However, this estimation of the Gaussian distribution's parameters is not optimal for images with low SNR. Besides, the fuzzy C-Mean algorithm is used for the segmentation of RGB images in [36]. The fuzzy membership is taken as basic belief assignment. Since the fuzzy membership can be interpreted rather as a particular plausibility function in the Dempster-Shafer evidence theory [38], it is improper to take the fuzzy membership as basic belief assignment.

## 1.3 Proposal

In this paper, the macro-structure of DEM is employed and its diffusion step is generalized to an intermediate step (I-step) as presented in Fig. 2. The likelihood function of Sanjay-Gopal *et al.* is chosen. The correlation between pixels, which are spatially far away from each other, is decoupled. Furthermore, the classical Gaussian mixture is replaced by a generalized mixture model, whose components are chosen from a Pearson system [39]. There is a set of eight types of distributions in a Pearson system. The components of the mixture model are no longer required to be of the same distribution type. Therefore, the generalized mixture model is more flexible to approximate the statistics of sonar data. In addition, we apply the Dempster-Shafer evidence theory based clustering technique in an I-step. The neighbors of a pixel are considered as evidence that support the hypotheses regarding the class label of this pixel.

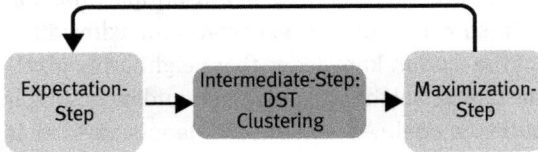

**Fig. 2.** There is an I-step inserted between the E- and M-step of the EM algorithm.

This paper is organized as follows. In Section 2, the maximum likelihood estimation, the Pearson system and EM algorithm are presented. The spatial dependency among pixels and the proposed EM approach using Dempster-Shafer evidence theory are explained in Section 3. Finally, numerical studies are carried out using SAS images in Section 4. Conclusions are drawn in Section 5.

## 2 Maximum likelihood estimation

Let $u_i$ denote the intensity of the $i$-th pixel in the image,

$$u_i = u_i + \epsilon_i,\tag{1}$$

where $u_i \in \mathfrak{U}$ denotes the intensity of pixel $i$ in the unknown noise-free image, $\epsilon_i$ is additive noise and $\mathfrak{U}$ is the set of all possible states of $u_i$. Let $\mathcal{L}$ be a set of labels with $|\mathcal{L}| = M_l$. The task of image segmentation is to assign to each $u_i$ a membership label $l_i \in \mathcal{L}$, cf. Fig. 3. For notational convenience, we denote the image as a vector $\mathbf{u} = \left(u_1, ..., u_i, ..., u_{N_u}\right)^T$, where $N_u$ is the number of pixels in the image,

$i \in \mathcal{I} = \{1, 2, ..., N_u\}$. Analogously, the corresponding labels are represented by $\mathbf{l} = (l_1, ..., l_i, ..., l_{N_u})^T$.

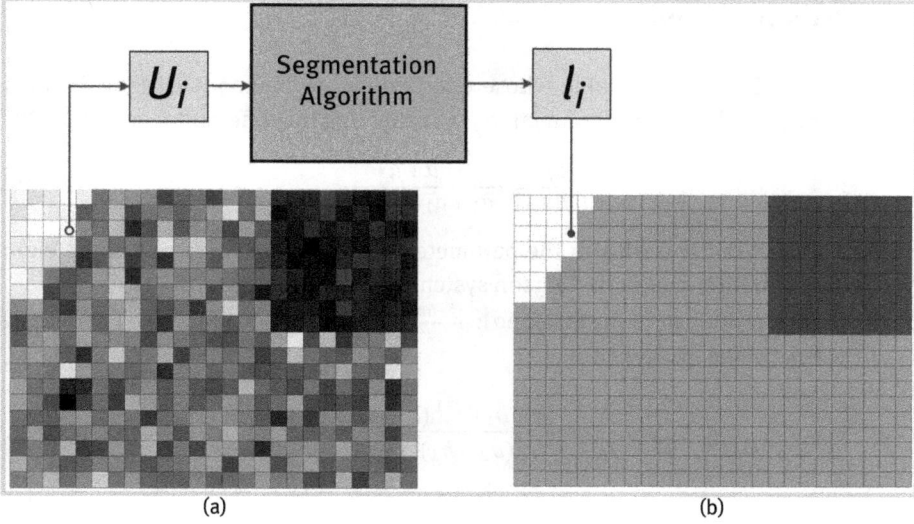

(a)                                                    (b)

**Fig. 3.** Illustration of image segmentation. (a) Image **u** and (b) Associated labels **l**.

Furthermore, an indicator vector $\mathbf{r}_i$:

$$\mathbf{r}_i = (r_{i,1}, ... r_{i,j}, ..., r_{i,M_l})^T \in \{\mathbf{e}_1, ..., \mathbf{e}_{M_l}\}$$

for $M_l = |\mathcal{L}|$ is introduced. Hence, we can define the probability

$$p(l_i = j) = p(\mathbf{r}_i = \mathbf{e}_j) = \pi_{i,j}, \tag{2}$$

where $\pi_{i,j}$ is a mixing coefficient with $0 \leq \pi_{i,j} \leq 1$, $\sum_{j=1}^{M_l} \pi_{i,j} = 1$ and $\mathbf{e}_j$ is an unit vector whose $j$-th component is 1. Let vector $\mathbf{\mathfrak{z}} = \left(\mathbf{u}^T, \mathbf{r}_1^T, ..., \mathbf{r}_{N_u}^T\right)^T$ be the complete data. Sanjay-Gopal *et al.* [40] propose the conditional probability density function

$$p(\mathbf{\mathfrak{z}}|\boldsymbol{\Phi}) = \prod_{i=1}^{N_u} \prod_{j=1}^{M_l} \left[\pi_{i,j} f_U(u_i|\boldsymbol{\psi}_j)\right]^{r_{i,j}}, \tag{3}$$

where $f_U$ is a probability density function, $\boldsymbol{\Phi} = \left(\boldsymbol{\Pi}^T, \boldsymbol{\Psi}^T\right)^T$ is the parameter vector with $\boldsymbol{\Pi} = \left(\boldsymbol{\pi}_1^T, ..., \boldsymbol{\pi}_{N_u}^T\right)^T$, $\boldsymbol{\Psi} = \left(\boldsymbol{\psi}_1^T, ..., \boldsymbol{\psi}_{M_l}^T\right)^T$, $\boldsymbol{\pi}_i = (\pi_{i,1}, ..., \pi_{i,M_l})^T$ and $\boldsymbol{\psi}_j$ denoting parameters for the probability density function of class $j$. In this paper, the $f_U$ is chosen from a Pearson system.

## 2.1 Pearson system

Let $U$ be a real random variable whose distribution can be modeled by a Pearson system. The probability density function $f(u)$ satisfying the differential equation [39]

$$\frac{1}{f}\frac{df}{du} = -\frac{a+u}{a_0 + a_1 u + a_2 u^2} \tag{4}$$

belongs subject to the setting of the parameters $a_0$, $a_1$, $a_2$ and $a$ to one of the eight possible distribution types of a Pearson system stated below.

1. $F_1$ (Beta distribution of the first kind): $\frac{a_1^2 - 4a_0 a_2}{a_2^2} > 0$ and $\frac{a_0}{a_2} < 0$ for $a_2 \neq 0$. The density function can be expressed as

$$f(u) = \begin{cases} \dfrac{1}{B(\tau_1, \tau_2)} \times \dfrac{(u-b_1)^{\tau_1-1}(b_2-u)^{\tau_2-1}}{(b_2-b_1)^{\tau_1+\tau_2-1}}, & \text{for } u \in [b_1, b_2] \\ 0, & \text{otherwise} \end{cases} \tag{5}$$

with

$$b_1 = -\frac{a_1}{2a_2} - 0.5\sqrt{\frac{a_1^2 - 4a_0 a_2}{a_2^2}}, \quad b_2 = -\frac{a_1}{2a_2} + 0.5\sqrt{\frac{a_1^2 - 4a_0 a_2}{a_2^2}},$$

$$\tau_1 = \frac{a+b_1}{a_2(b_2-b_1)} + 1, \quad \tau_2 = -\frac{a+b_2}{a_2(b_2-b_1)} + 1,$$

and where $B(\tau_1, \tau_2)$ denotes the beta function.

2. $F_2$ (Type II Distribution): It is a particular case of $F_1$ with $\tau_1 = \tau_2$, and the density function is as follows,

$$f(u) = \begin{cases} \dfrac{1}{B(\tau, \tau)} \times \dfrac{(u-b_1)^{\tau-1}(b_2-u)^{\tau-1}}{(b_2-b_1)^{2\tau-1}}, & \text{for } u \in [b_1, b_2] \\ 0, & \text{otherwise} \end{cases} \tag{6}$$

with $\tau = \frac{a+b_1}{a_2(b_2-b_1)} + 1, b_1 = -a - 0.5\sqrt{\frac{a_1^2 - 4a_0 a_2}{a_2^2}}$, and $b_2 = -a + 0.5\sqrt{\frac{a_1^2 - 4a_0 a_2}{a_2^2}}$.

3. $F_3$ (Gamma distribution): $a_2 = 0$ (and $a_1 \neq 0$). In this case, the density function is

$$f(u) = \begin{cases} \dfrac{1}{\tau_1 \Gamma(\tau_2)} \left(\dfrac{u-\tau_3}{\tau_1}\right)^{\tau_2-1} \exp\left(-\dfrac{u-\tau_3}{\tau_1}\right), & \text{for } u \geq \tau_3 \\ 0, & \text{otherwise} \end{cases} \tag{7}$$

with $\tau_1 = a_1$, $\tau_2 = \dfrac{1}{a_1}\left(\dfrac{a_0}{a_1} - a\right) + 1$, $\tau_3 = -\dfrac{a_0}{a_1}$ and $\Gamma$ denotes the gamma function.

4. $F_4$ (Type IV distribution): $a_1^2 - 4a_0a_2 < 0$. The associated distribution density function is

$$f(u) = \mathfrak{N}_1 \left[\tau_1 + a_2(u + \tau_2)^2\right]^{-(1/2a_2)} \tag{8}$$
$$\exp\left(-\frac{a - \tau_2}{\sqrt{\tau_1 a_2}} \arctan\left(\sqrt{\frac{a_2}{\tau_1}}(u + \tau_2)\right)\right),$$

with $\tau_1 = a_0 - \dfrac{a_1^2}{4a_2}$, $\tau_2 = \dfrac{a_1}{2a_2}$, and where $\mathfrak{N}_1$ has to be chosen such that $\int_{\mathcal{R}} f(u)du = 1$.

5. $F_5$ (Type V distribution): $a_1^2 = 4a_0a_2$. The associated distribution density function is

$$f(u) = \begin{cases} \dfrac{\tau_1}{\Gamma(\tau_2)}\left[\tau_1\left(u + \dfrac{a_1}{2a_2}\right)\right]^{-\tau_2-1} \exp\left[\dfrac{-2}{\tau_1\left(u + \frac{a_1}{2a_2}\right)}\right], & \text{for } u \geq -\dfrac{a_1}{2a_2} \\ 0, & \text{otherwise} \end{cases} \tag{9}$$

with $\tau_1 = \dfrac{a_2}{a - \frac{a_1}{2a_2}}$ and $\tau_2 = \dfrac{1}{a_2} - 1$.

6. $F_6$ (Beta distribution of the second kind): $\dfrac{a_1^2 - 4a_0a_2}{a_2^2} \geq 0$ and $\dfrac{a_0}{a_2} > 0$. The associated distribution density function is

$$f(u) = \begin{cases} \dfrac{\tau_4^{\tau_2}}{B(\tau_1, \tau_2)} \times \dfrac{(u - \tau_3)^{\tau_1-1}}{(u - (\tau_3 - \tau_4))^{\tau_1+\tau_2}}, & \text{for } u \geq \tau_3 \\ 0, & \text{otherwise} \end{cases} \tag{10}$$

with (for $a_2 \neq 0$):

$$\tau_1 = -\frac{a - \frac{1}{2a_2}\left(a_1 - \sqrt{a_1^2 - 4a_0a_2}\right)}{\sqrt{a_1^2 - 4a_0a_2}} + 1, \quad \tau_2 = \frac{1}{a_2} - 1$$

$$\tau_3 = -\frac{1}{2a_2}\left(a_1 - \sqrt{a_1^2 - 4a_0a_2}\right), \quad \tau_4 = \sqrt{\frac{a_1^2 - 4a_0a_2}{a_2^2}}$$

7. $F_7$ (Type VII distribution): $a_1 = a = 0$, $a_0 > 0$, and $a_2 > 0$. The corresponding density function is given as [41]

$$f(u) = \frac{1}{\mathcal{B}(0.5, \tau_1 - 0.5)} \times \frac{\tau_2^{2\tau_1-1}}{(2\tau_1)^{\tau_1}}\left(\frac{\tau_2^2}{2\tau_1}\left[1 + \left(\frac{u}{\tau_2}\right)^2\right]\right)^{-\tau_1}, \tag{11}$$

where $\tau_1 = \dfrac{1}{2a_2}, \tau_2 = \sqrt{2\tau_1 a_0}$, and:

$$\mathcal{B}(\tau_3, \tau_4) = \int\limits_0^\infty \frac{u^{\tau_3 - 1}}{(1+u)^{(\tau_3 + \tau_4)}} \, du$$

8. $F_8$ (Gaussian distribution): $a_1 = a_2 = 0$. Thus, the associated density function is

$$f(u) = \frac{1}{\sqrt{2\pi\sigma^2}} \exp\left[-\frac{(u-\mu)^2}{2\sigma^2}\right] \tag{12}$$

with $\mu = -a$ and $\sigma^2 = a_0$.

As summarized above, the determination of the distribution type is dependent on the values of the parameters $a, a_0, a_1$ and $a_2$. However, they are usually unknown *a priori*. Johnson *et al.* demonstrated that it is possible to express $a, a_0, a_1$ and $a_2$ in terms of central moments as follows [42],

$$a = \frac{(\mathfrak{s}_2 + 3)\sqrt{\mathfrak{s}_1 \zeta_2}}{10\mathfrak{s}_2 - 12\mathfrak{s}_1 - 18} - \mu, \tag{13}$$

$$a_0 = \frac{\zeta_2(4\mathfrak{s}_2 - 3\mathfrak{s}_1) - \mu(\mathfrak{s}_2 + 3)\sqrt{\mathfrak{s}_1 \zeta_2} + \mu^2(2\mathfrak{s}_2 - 3\mathfrak{s}_1 - 6)}{10\mathfrak{s}_2 - 12\mathfrak{s}_1 - 18}, \tag{14}$$

$$a_1 = \frac{(\mathfrak{s}_2 + 3)\sqrt{\mathfrak{s}_1 \zeta_2} - 2\mu(2\mathfrak{s}_2 - 3\mathfrak{s}_1 - 6)}{10\mathfrak{s}_2 - 12\mathfrak{s}_1 - 18}, \tag{15}$$

$$a_2 = \frac{2\mathfrak{s}_2 - 3\mathfrak{s}_1 - 6}{10\mathfrak{s}_2 - 12\mathfrak{s}_1 - 18}, \tag{16}$$

where $\mu = E[U]$, $\zeta_n = E\left[(U - \mu)^n\right]$ for $n = 2, 3, 4$, as well as $\mathfrak{s}_1 = \frac{(\zeta_3)^2}{(\zeta_2)^3}$ and $\mathfrak{s}_2 = \frac{\zeta_4}{(\zeta_2)^2}$. Hence, the classification of the distribution type, which was based on the setting of $a, a_0, a_1$ and $a_2$, can be done via the moments. The advantage of this conversion is that in practical applications the central moments can be estimated from the data. Based on the moments, the rule can be reformulated as follows,

$$\begin{cases} f \in F_1, & \text{for } \lambda < 0, \\ f \in F_2, & \text{for } \mathfrak{s}_1 = 0 \text{ and } \mathfrak{s}_2 < 3, \\ f \in F_3, & \text{for } 2\mathfrak{s}_2 - 3\mathfrak{s}_1 - 6 = 0, \\ f \in F_4, & \text{for } 0 < \lambda < 1, \\ f \in F_5, & \text{for } \lambda = 1, \\ f \in F_6, & \text{for } \lambda > 1, \\ f \in F_7, & \text{for } \mathfrak{s}_1 = 0 \text{ and } \mathfrak{s}_2 > 3, \\ f \in F_8, & \text{for } \mathfrak{s}_1 = 0 \text{ and } \mathfrak{s}_2 = 3, \end{cases} \tag{17}$$

where $\lambda = \dfrac{\mathfrak{s}_1(\mathfrak{s}_2 + 3)^2}{4(4\mathfrak{s}_2 - 3\mathfrak{s}_1)(2\mathfrak{s}_2 - 3\mathfrak{s}_1)(2\mathfrak{s}_2 - 3\mathfrak{s}_1 - 6)}.$

## 2.2 Expectation-maximization with generalized mixture model

The EM algorithm is an elegant method to maximize the likelihood function in (3). It iterates itself between E-step and M-step. The E-step takes a conditional expectation as follows,

$$Q\left(\boldsymbol{\Phi}|\boldsymbol{\Phi}^{(k)}\right) = E\left[\ln\left(p(\boldsymbol{\mathfrak{z}}|\boldsymbol{\Phi})\right)|\boldsymbol{\Phi}^{(k)},\mathbf{U}=\mathbf{u}\right], \tag{18}$$

$$= E\left[\sum_{i=1}^{N_u}\sum_{j=1}^{M_l}r_{i,j}\left(\ln\pi_{i,j}+\ln f_U(u_i|\boldsymbol{\psi}_j)\right)|\boldsymbol{\Phi}^{(k)},\mathbf{U}=\mathbf{u}\right],$$

where $\mathbf{U} = \left(U_1, ..., U_{N_u}\right)^T$, and $\boldsymbol{\Phi}^{(k)}$ denotes the parameters obtained in the $k$-th iteration. Similar as derived in the Appendix C of [40], we have

$$E\left[r_{i,j}|\boldsymbol{\Phi}^{(k)}\right] = \frac{\pi_{i,j}^{(k)}f_U\left(u_i|\boldsymbol{\psi}_j^{(k)}\right)}{\sum_{m=1}^{M_l}\pi_{i,m}^{(k)}f_U\left(u_i|\boldsymbol{\psi}_m^{(k)}\right)} = w_{i,j}^{(k)}. \tag{19}$$

In the M-step, the $\boldsymbol{\Phi}^{(k)}$ is updated with

$$\boldsymbol{\Phi}^{(k+1)} = \arg\max_{\boldsymbol{\Phi}} Q\left(\boldsymbol{\Phi}|\boldsymbol{\Phi}^{(k)}\right). \tag{20}$$

In the mixture model, pixels are considered as conditionally independent. Thus, (20) can be maximized pixel by pixel. For pixel $i$, conditioned on $\sum_{j=1}^{M_l}\pi_{i,j} = 1$, a Lagrange multiplier $\mathfrak{L}$ is introduced,

$$\Lambda = \sum_{j=1}^{M_l}w_{i,j}^{(k)}\left[\ln\pi_{i,j}+f_U\left(u_i|\boldsymbol{\psi}^{(k)}\right)\right] + \mathfrak{L}\left(\sum_{j=1}^{M_l}\pi_{i,j}-1\right). \tag{21}$$

Through $\dfrac{\partial\Lambda}{\partial\pi_{i,j}} = 0$ for $j = 1, .., M_l$ we get

$$\frac{w_{i,j}^{(k)}}{\pi_{i,j}} + \mathfrak{L} = 0 \quad\text{and}\quad \frac{\sum_{j=1}^{M_l}w_{i,j}^{(k)}}{\mathfrak{L}} = -1, \quad j = 1, ..., M_l. \tag{22}$$

Solving (22), we get the update of $\pi_{i,j}$,

$$\pi_{i,j}^{(k+1)} = w_{i,j}^{(k)}, \tag{23}$$

and the mean and the central moments are updated as follows,

$$
\mu_j^{(k+1)} = \frac{\sum_{i=1}^{N_u} u_i \pi_{i,j}^{(k+1)}}{\sum_{i=1}^{N_u} \pi_{i,j}^{(k+1)}} = \frac{\sum_{i=1}^{N_u} u_i w_{i,j}^{(k)}}{\sum_{i=1}^{N_u} w_{i,j}^{(k)}} \tag{24}
$$

$$
\zeta_{n,j}^{(k+1)} = \frac{\sum_{i=1}^{N} \left(u_i - \mu_j^{(k+1)}\right)^n \pi_{i,j}^{(k+1)}}{\sum_{i=1}^{N} \pi_{i,j}^{(k+1)}} = \frac{\sum_{i=1}^{N} \left(u_i - \mu_j^{(k+1)}\right)^n w_{i,j}^{(k)}}{\sum_{i=1}^{N} w_{i,j}^{(k)}} \tag{25}
$$

for $n = 2, 3, 4$, and where $\mu_j$ and $\zeta_{n,j}$ are the mean value and the $n$-th central moment of the pixels belonging to class $j$, respectively. With the results in (24) and (25), $\mathfrak{s}_1$, $\mathfrak{s}_2$ and $\lambda$ can be obtained. Accordingly, the distribution types of $f_U$ and their associated parameters can be determined as described in Section 2.1 for the next EM iteration.

# 3 Spatial dependency among pixels

## 3.1 Dempster-Shafer evidence theory

Although it was completed more than forty years ago, many engineers are not familiar with the Dempster-Shafer evidence theory (DST). Hence, it is meaningful to review some basics of the DST before introducing our modeling. The DST allows one to combine the information from different pieces of evidence and arrive at a degree of belief, which takes into account all the available evidence. In DST [43], the finite set of class indices $\mathcal{L}$ is called the frame of discernment. One's total belief induced by a piece of evidence concerning $\mathcal{L}$ is assumed to be partitioned into various portions, each of which is assigned to a subset of $\mathcal{L}$. The function $\mathfrak{b}: 2^{\mathcal{L}} \to [0, 1]$ describing this belief portion assignment and satisfying the following conditions,

$$
\mathfrak{b}(\emptyset) = 0 \quad \text{and} \quad \sum_{\Theta \subseteq \mathcal{L}} \mathfrak{b}(\Theta) = 1, \tag{26}
$$

is called basic belief assignment (bba). The quantity $\mathfrak{b}(\Theta)$ is committed exactly to $\Theta$, and not to any of its subsets. Every $\Theta \in 2^{\mathcal{L}}$ that satisfies $\mathfrak{b}(\Theta) > 0$ is called a focal element of the bba. From the bba, the belief function is defined,

$$
Bel(\Theta) = \sum_{\theta \subseteq \Theta} \mathfrak{b}(\theta). \tag{27}
$$

The quantity $Bel(\Theta)$ represents the total belief committed to the hypothesis $\Theta$. It can easily be verified [44] that the $Bel(\Theta)$ and the $Bel(\bar{\Theta})$ do not necessarily sum to 1. This is a major difference to probability theory. Moreover, another quantity $Pl(\Theta) = 1 - Bel(\bar{\Theta})$ called plausibility is defined to describe the extent to which one fails to doubt in $\Theta$. Hence, the probability of hypothesis $\Theta$ is bounded by $Bel$ and $Pl$, $Bel(\Theta) \le P(\Theta) \le Pl(\Theta), \forall \Theta \subseteq \mathcal{L}$.

The Dempster's rule is a mathematical operation used to combine two bba's induced by different pieces of evidence, $b_1$ and $b_2$,

$$b_{1 \oplus 2}(\Theta) = \frac{\sum\limits_{\Theta_1 \cap \Theta_2 = \Theta} b_1(\Theta_1) b_1(\Theta_2)}{1 - \sum\limits_{\Theta_1 \cap \Theta_2 = \emptyset} b_1(\Theta_1) b_1(\Theta_2)}, \tag{28}$$

where $\Theta, \Theta_1, \Theta_2 \in 2^{\mathcal{L}}$. Since the Dempster's rule is commutative and associative, the bba's of diverse evidence can be combined sequentially in any arrangement. The decision-making of DST is still open. There exists an interval of probabilities bounded by $Bel$ and $Pl$. Consequently, simple hypotheses can no longer be ranked according to their probabilities. Over the last thirty years, many proposals have been made to conquer this ambiguity on probabilities. In this paper, we use the well-known pignistic probability [45] proposed by P. Smets,

$$BetP(\Theta) = \sum\limits_{\Theta' \subseteq \mathcal{L}} b(\Theta') \frac{|\Theta \cap \Theta'|}{|\Theta'|}. \tag{29}$$

## 3.2 Dempster-Shafer evidence theory assisted EM segmentation

In the framework of DST, we model the neighbors as pieces of evidence. They provide support to the hypotheses that the pixel of interest belongs to the same classes of these neighbors.

As depicted in Fig. 4, it is a second order neighborhood of the pixel of interest, i.e. pixel $i$. This neighborhood is denoted as $\mathcal{N}_i$. All of its neighbors are labeled, and can be used as evidence. The amount of support provided by a neighbor $\eta$ to the hypothesis that pixel $i$ is assigned with the same label as pixel $\eta$ relies on the difference between $u_i$ and the average of all the $u_{i'}$ with $l_{i'} = l_\eta$. Hence, the variation caused by the noise contained in the observation of the neighbors can be minimized. A small difference in the pixel intensities should indicate a great amount of support. If a neighbor $\eta \in \mathcal{N}_i$ belongs to class $l_\eta \in \mathcal{L}$, its bba is given as:

$$b(\Delta) = \begin{cases} \vartheta_\eta v_\eta, & \text{if } \Delta = \{l_\eta\}, \\ 1 - \vartheta_\eta v_\eta, & \text{if } \Delta = \mathcal{L}, \end{cases} \tag{30}$$

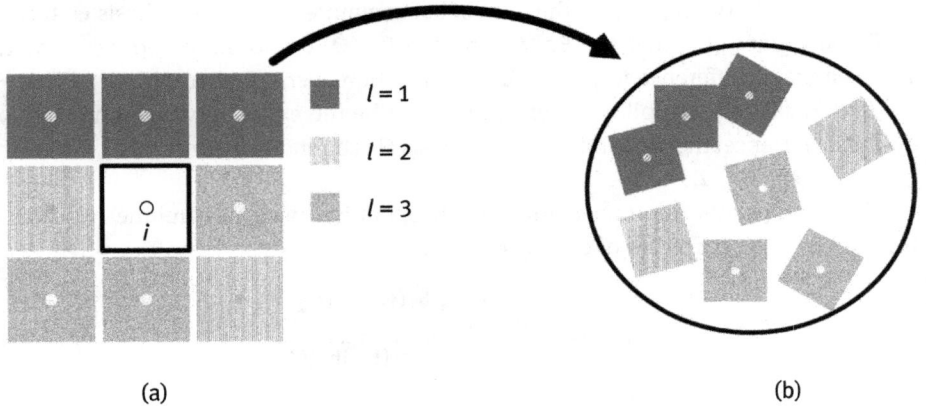

(a)                                                                                          (b)

**Fig. 4.** (a): The neighborhood configuration of pixel $i$. (b): The evidence pool.

where the $\vartheta_\eta$ and $\nu_\eta$ are determined by

$$\vartheta_\eta = \frac{\exp(-\gamma_1 |u_\eta - \nu_i|)}{\max\limits_{\eta' \in \mathcal{N}_i} \exp(-\gamma_1 |u_{\eta'} - \nu_i|)}, \tag{31}$$

$$\nu_\eta = \exp\left(-\gamma_2 \frac{|u_i - \mu_{l_\eta}|}{\sigma_{l_\eta}}\right), \tag{32}$$

where $\mu_{l_\eta}$ and $\sigma_{l_\eta}$ are the mean value and standard deviation of class $l_\eta$, $\nu_i$ is the median of the pixel intensity of $\mathcal{N}_i$, and $\gamma_1, \gamma_2$ are positive constants. The choice of $\gamma_1$ and $\gamma_2$ will be motivated in Section 4. The $\nu_\eta$ denotes the total belief portion which is able to be provided by the pixel $\eta \in \mathcal{N}_i$, and the $\vartheta_\eta$ evaluates the quality of the evidence. This quality evaluation is based on the assumption that the information supplied by an outlier is less plausible.

There is an effective combination scheme for the simple bba derived from Dempster's rule by Denoeux *et al.* in [44],

$$\mathfrak{b}_{\text{total}}(\{l\}) = \frac{\mathfrak{b}^{(l)}(\{l\}) \prod\limits_{l' \neq l} \mathfrak{b}^{(l')}(\mathcal{L})}{\mathfrak{K}}, \tag{33}$$

$$\mathfrak{b}_{\text{total}}(\mathcal{L}) = \frac{\prod\limits_{l \in \mathcal{L}} \mathfrak{b}^{(l)}(\mathcal{L})}{\mathfrak{K}}, \tag{34}$$

where $\mathfrak{b}^{(l)}$ and the normalized factor $\mathfrak{K}$ are given by

$$\mathfrak{b}^{(l)}(\{l\}) = 1 - \prod\limits_{\eta \in \mathcal{N}_i^l} (1 - \mathfrak{b}_\eta(\{l\})), \tag{35}$$

$$b^{(l)}(\mathcal{L}) = \prod_{\eta \in \mathcal{N}_i^l} \left(1 - b_\eta(\{l\})\right), \tag{36}$$

$$\mathfrak{K} = \sum_{l \in \mathcal{L}} \prod_{l' \neq l} b^{(l')}(\mathcal{L}) + (1 - |\mathcal{L}|) \prod_{l \in \mathcal{L}} b^{(l)}(\mathcal{L}) \tag{37}$$

with $\mathcal{N}_i^l \subseteq \mathcal{N}_i$ denoting the set of neighbors in $\mathcal{N}_i$ belonging to the class $l \in \mathcal{L}$ and $b_\eta$ the bba associated with the neighbor $\eta$. We choose the well known pignistic probability [45] for the sake of decision-making. Due to the fact that focals of $b_{\text{total}}$ are either singletons $\{l\} \subset \mathcal{L}$ or $\mathcal{L}$ itself, the results obtained from the pignistic level is identical to those from the bba function. Thus, the decision-making for pixel $i$ is given by

$$l_i = \arg\max_{l \in \mathcal{L}} b_{\text{total},i}(\{l\}), \tag{38}$$

where $b_{\text{total},i}$ is the combined bba associated with pixel $i$. Finally, we should turn the label information into $w_{i,j}$ by

$$\bar{w}_{i,j} = \begin{cases} 1, & l_i = j, \\ 0, & l_i \neq j. \end{cases} \tag{39}$$

Hence, the proposed method called E-DS-M can be summarized as follows,
1. Initialization: gamma mixture is chosen for the first iteration
2. Run E-step with the help of (19), and obtain $\left\{w_{i,j}^{(k)}\right\}$
3. Perform a hard decision on $\left\{w_{i,j}^{(k)}\right\}$, then get $\left\{l_i^{(k)}\right\}$
4. Determine the bba as shown in (30), (31) and (32)
5. Combine the bba's with the assistance of (33), (34), (35) and (36)
6. Determine the $l_i$ and $\left\{\bar{w}_{i,j}^{(k)}\right\}$ by (38) and (39), respectively
7. Forward $\left\{\bar{w}_{i,j}^{(k)}\right\}$ to M-step, substitute the $w_{i,j}^{(k)}$ with $\bar{w}_{i,j}^{(k)}$ in (24) and (25), and estimate the central moments of each class, $\mu_j^{(k+1)}$ and $\zeta_{n,j}^{(k+1)}$ using (24) and (25)
8. Determine the types of $f_U$ in (3) with the help of (17)
9. Return to **Step 2** until results converge or the number of maximum iterations is reached

# 4 Numerical study

Numerical tests are carried out on both real SAS data and synthetic data. The ripple-like sediment is a great challenge for sonar image segmentation. Owing to the high cost of sea trials, the availability of real sonar data is limited. We have only the SAS data that is obtained from sea trails launched on flat sediments. Thus, we simulate the SAS data with ripple-like sediment to verify the reliability of E-DS-M. It is found in our study that E-DS-M can provide almost perfect results on ripple-like sediments.

The performance gain against the methods in the literature can be easily observed. Therefore, there is no necessity to use additional measure for the evaluation of the result obtained from synthetic data. In contrast, due to the complexity of real SAS images, a quantitative measure dedicated to image segmentation is required for the performance evaluation.

The results of E-DS-M compared to those of the methods from the literature: MAP estimation solved by ICM algorithm (MAP-ISO) [28], DEM [32], and the method proposed by Reed *et al.* (MAP-Reed)[3].

## 4.1 Experiments on real SAS images

We employ the variation of information (VI) [46] to evaluate the segmentation results. VI, $I_{VI}$, provides us the measure on dissimilarity between segmentation result and the groundtruth. The $I_{VI} = 0$ means that the segmentation result is perfectly identical to the groundtruth. The SAS images used for numerical study are given in Fig. 5. Their dimensions are $100 \times 100$ pixels.

In order to visualize the impact of $\gamma_1$ and $\gamma_2$, we vary them to reveal how the E-DS-M reacts to the tuning of parameters. We compute the $I_{VI}$ of all the test images in Fig. 5 and present the averages of $I_{VI}$ over the eight images in Fig. 6.

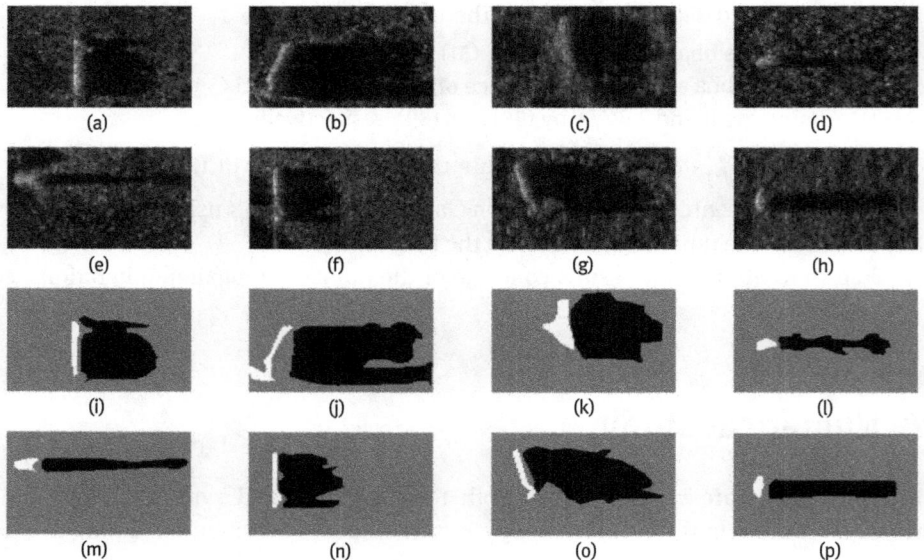

**Fig. 5.** The SAS image and the associated ground truth. (a)–(h) are SAS images and their groundtruths are in (i)–(p).

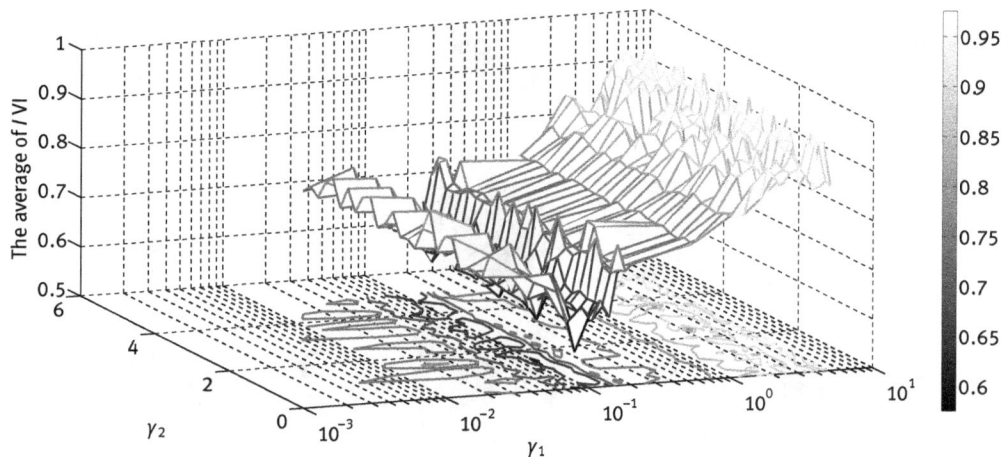

**Fig. 6.** The averages of the $I_{VI}$ over the 8 test images in Fig. 5.

Obviously, although the variation of the $\gamma_2$ in (32) has some influence on the performance of image segmentation, it is neither significant nor definite. In opposite, the performance of image segmentation is highly dependent on the setting of $\gamma_1$ in (31). As the $\gamma_1$ grows, more neighbors are recognized as outliers and their support to the corresponding hypotheses is suppressed. The consequence is that the useful information embodied in the neighbors could be ignored and the segmentation results of the E-DS-M are impaired. There is a significant performance degradation around $\gamma_1 = 0.2$. According to the results in the Fig. 6, the E-DS-M has a satisfying performance when the $\gamma_1$ is around 0.1. We find that the optimal parameter setting in this test is $\gamma_1 = 0.1$ and $\gamma_2 = 1.4$. Using this optimal setting, the $I_{VI}$ of individual test images are presented in Fig. 7. For the comparison, the $I_{VI}$ of MAP-ISO, DEM and MAP-Reed

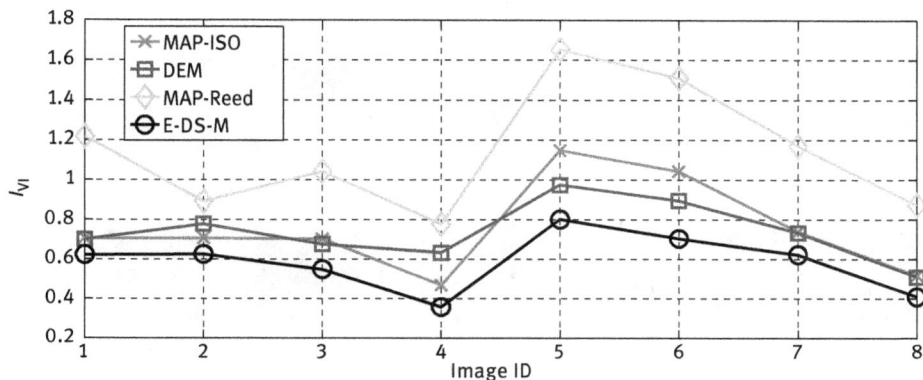

**Fig. 7.** The $I_{VI}$ of individual test images. The x-axis denotes ID of the image (a)-(h) in Fig. 5.

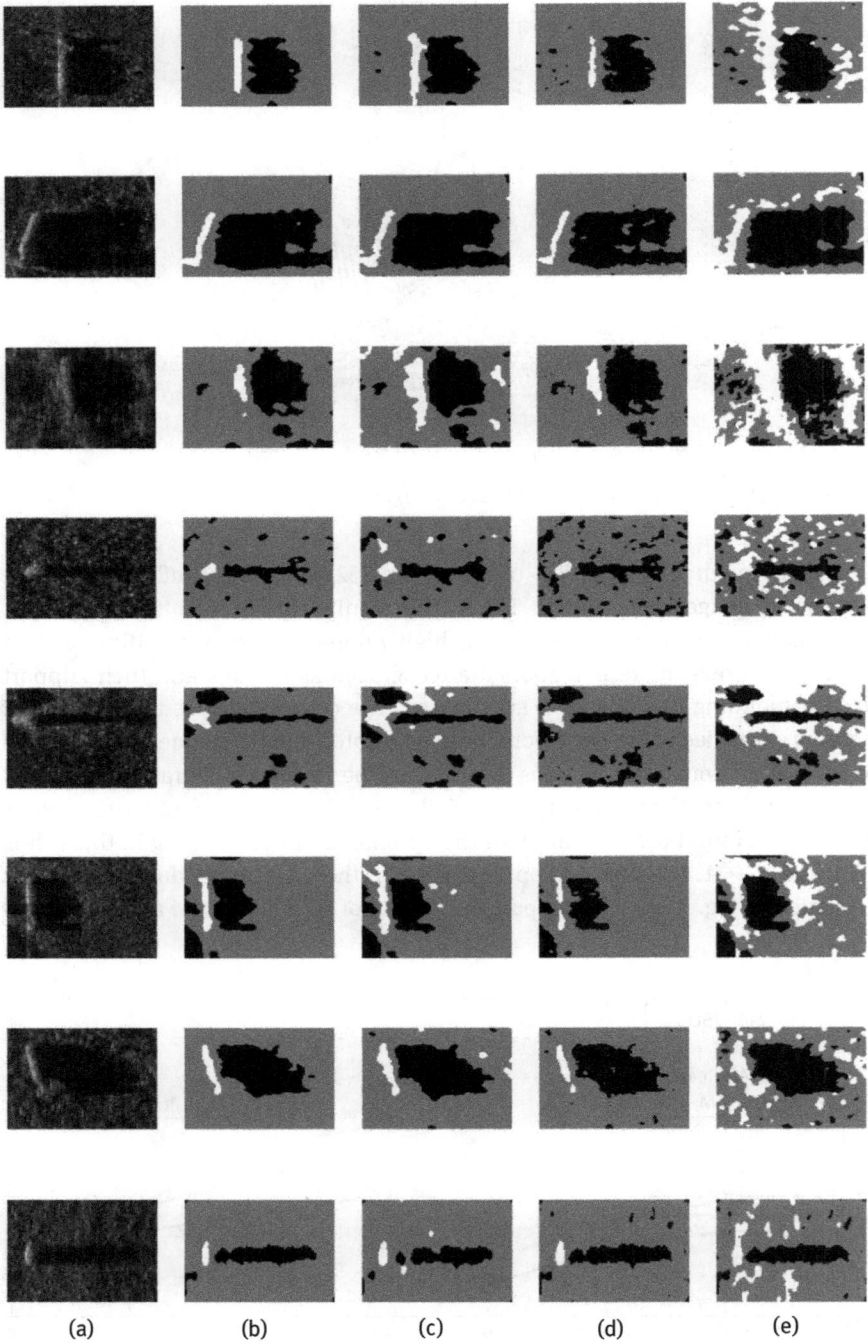

Fig. 8. Examples of the segmentation results. Column (a) presents the sonar images, in column (b) up to column (e) there are segmentation results obtained by the methods E-DS-M, MAP-ISO, DEM and MAP-Reed respectively.

are also shown. It is apparent that E-DS-M can outperform the other three methods. Finally, we visualize this comparison of segmentation results in Fig. 8. It is apparent in Fig. 8 that the results given by E-DS-M can provide more precise segmentation results with less mislabeled pixels than other methods.

## 4.2 Experiments on synthetic images

The performance of E-DS-M on SAS images with ripple-like sediments are studied in this subsection. There is a synthetic image containing cylinder mines for image segmentation, cf. Fig. 9(a). It is generated as described in [47]. Its dimension is $300 \times 300$ pixels. The same parameter setting for $\gamma_1$ and $\gamma_2$ as in Fig. 8 is applied to the test on synthetic images. The results are shown in Fig. 9(c)-(f).

Comparing the result of E-DS-M with those of MAP-ISO, DEM and MAP-Reed, it can be observed that E-DS-M can suppress the influence of a ripple-like sediment very well. The segmentation result is almost identical to the ground truth. Thus, it is verified that E-DS-M is also reliable when objects are lying on ripple-like sediments.

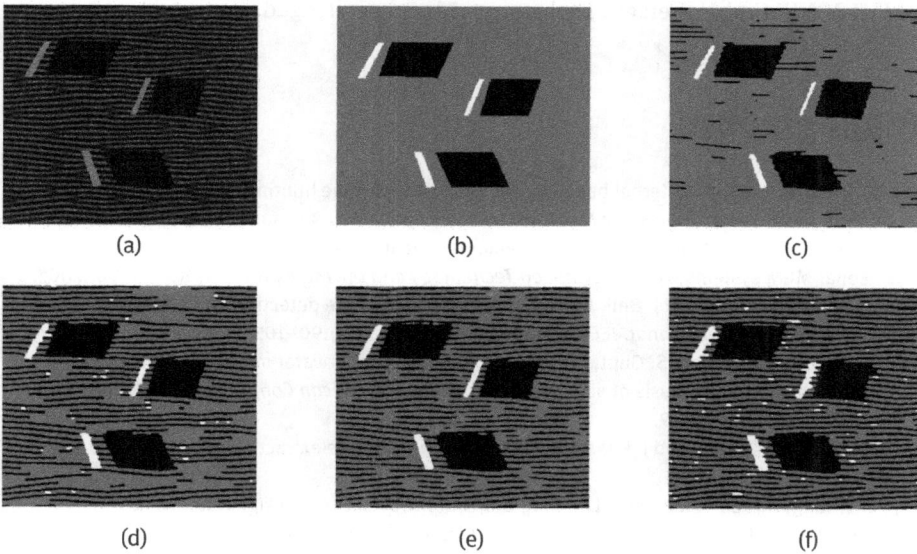

(a)          (b)          (c)

(d)          (e)          (f)

**Fig. 9.** The numerical test on synthetic image: (a) synthetic image with ripple-like sediment, (b) ground truth, (c)-(d) provide the segmentation results given by E-DS-M, MAP-ISO, DEM and MAP-Reed, respectively.

# 5 Conclusion

A generalized expectation-maximization approach assisted by evidence theory for image segmentation is developed. The expectation-maximization approach of Delignon *et al.* using Pearson system is extended by substituting its mixture model with the one proposed by Sanjay-Gopal *et al.* A Dempster-Shafer evidence theory based clustering method is incorporated to consider the dependency among neighboring pixels. A simple belief structure is proposed to catch the belief portion provided by evidence in the neighborhood. It considers not only the amount of the belief that the evidence can provide but also the quality of the evidence itself. Compared with the methods in the literature, the proposed approach can considerably enhance the quality of the segmentation results. With the help of a quantitative analysis, optimal settings for the parameters involved in the construction of the belief structure are found. It is revealed that only the parameter that is involved in the evaluation of the quality of evidence is decisive. The other one has less influence. Thus, for the sake of simplicity, the number of parameters can be reduced to one in practical applications.

**Acknowledgment:** All the real SAS images used in this paper are provided by ATLAS ELEKTRONIK GmbH Bremen. Their support is acknowledged.

# Bibliography

[1]   J. Behn and D. Kraus. Kernel based target detection for mine hunting in sonar images. 4th *Int. Conf. on Underwater Acoustic Measurements: Technologies & Results*, :71–77, 2011.

[2]   Y. Petillot, S. Reed and E. Coiras. An augmented reality solution for evaluating underwater sonar MCM systems. 7th *Int. Symp. on Technology and the Mine Problem*, 1(1):83–90, 2007.

[3]   S. Reed, Y. Petillot and J. Bell. An automatic approach to the detection and extraction of mine features in sidescan sonar. *IEEE J. Oceanic Engeering*, 28(1):90–105, 2003.

[4]   C. Rao, K. Mukherjee, S. Gupta, A. Ray and S. Phoha. Underwater mine detection using symbolic pattern analysis of sidescan sonar images. *American Control Conf. (ACC'09)*, pages :5416–5421, June 2009.

[5]   M. Yang, K. Kpalma and J. Ronsin. A survey of shape feature extraction techniques. *Pattern Recognition*, 2008.

[6]   Y.Y. Tang, B.F. Li, H. Ma and J. Lin. Ring-projection-wavelet-fractal signatures: a novel approach to feature extraction. *IEEE Trans. Circuits Systems II: Analog Digital Signal Process*, 45(8):1130–1134, 1998.

[7]   J. S. Weszka, C. R. Dyer and A. Rosenfeld. A comparative study of texture measures for terrain classification. *IEEE Trans. Systems Man & Cybernetics*, 6(4):269–285, 1976.

[8]   M. Amadasun and R. King. Textural features corresponding to textural properties. *IEEE Trans. Systems, Man & Cybernetics*, 19(5):1264–1274, 1989.

[9]   Howard H. Yang and J. Moody. Feature selection based on joint mutual information. *Internat. ICSC Symp. on Advances in Intelligent Data Analysis (AIDA'99)*, :22–25, 1999.

[10] H. Peng, F. Long and C. Ding. Feature selection based on mutual information criteria of max-dependency, max-relevance, and min-redundancy. *IEEE Trans. Pattern Analysis & Machine Intelligence*, 27(8):1226–1238, 2005.

[11] N. Kwak and C. H. Choi. Input feature selection for classification problems. *IEEE Trans. Neural Networks*, 13(1):143–159, January 2002.

[12] T. Fei, D. Kraus and Paula Berkel. A new idea on feature selection and its application to the underwater object recognition. 11th *Europe Conf. Underwater Acoustics (ECUA 2012)*, :52–59, 2012.

[13] J. E. Goin. Classification bias of the $k$-nearest neighbor algorithm. *IEEE Trans. Pattern Analysis & Machine Intelligence*, 6(3):379–381, 1984.

[14] J. M. Keller, M. R. Gray and J. A. J. Givens. Fuzzy $k$-nearest neighbor algorithm. *IEEE Trans. Systems Man & Cybernetics*, 15(4):580–584, 1985.

[15] J.-P. Vert, K. Tsuda and B. Schölkopf. A primer on kernel methods. *Kernel Methods in Computational Biology* MIT Press, :35–70, 2004.

[16] G. P. Zhang. Neural networks for classification: a survey. *IEEE Trans. Systems, Man & Cybernetics, Part C: Applications and Reviews*, 30(4):451–462, 2000.

[17] D. F. Specht. Probabilistic neural networks. *Neural Networks*, 3(1):109–118, 1990.

[18] P. Langley and S. Sage. Induction of selective Bayesian classifiers. 10th *Annual Conf. Uncertainty in Artificial Intelligence (UAI'94)*, :399–406, 1994.

[19] N. Otsu. A threshold selection method from gray-level histograms. *IEEE Trans. Systems, Man & Cyberneticz*, 9:62–66, 1979.

[20] T. Pun. A new method for grey-level picture thresholding using the entropy of the histogram. *Signal Processing*, 2:223–237, 1980.

[21] P. K. Sahoo, D. W. Slaaf, and T.A. Albert. Threshold selection using a minimal histogram entropy difference. *Optical Engineering*, 36:1976–1981, 1997.

[22] P. K. Sahoo, S. Soltani, A. K. C. Wong, and Y. C. Chen. A survey of thresholding techniques. *Computer, Vision & Graphical Image Processing*, 41:233–260, 1998.

[23] E. W. Forgy. Cluster analysis of multivariate data: efficiency vs interpretability of classifications. *Biometrics*, 21:768–769, 1965.

[24] M. Kass, A. Witkin, and D. Terzopoulos. Snakes: Active contour models. *Int. J. Computer Vision*, 23:321–331, 1988.

[25] M. Lianantonakis and Y. R. Petillot. Sidescan sonar segmentation using active contours and level set methods. *Oceans - Europe*, 1:719–724, 2005.

[26] J. Besag. On the statistical analysis of dirty pictures. *J. R. Stat. Soc. Series B (Methodological)*, 48(3):259–302, 1986.

[27] M. Mignotte, C. Collet, P. Perez and P. Bouthemy. Sonar image segmentation using an unsupervised hierarchical MRF model. *IEEE Trans. Image Processing*, 9:1216–1231, 1998.

[28] O. Demirkaya, M. H. Asyali and P. Sahoo. *Image processing with MATLAB : applications in medicine and biology*. CRC Press, 6000 Broken Sound Parkway NW, Suite 300, Boca Raton, FL 33487-2742, 2009.

[29] R. Kindermann and J. L. Snell. *Markov Random Fields and Their Applications*. American Mathematical Society, 1980.

[30] A. P. Dempster, N. M. Laird and D. B. Rubin. Maximum likelihood from incomplete data via the EM algorithm. *J. R. Stat. Soc. Series B (Methodological)*, 39(1):1–38, 1977.

[31] J. Zhang, J.W. Modestino and D.A. Langan. Maximum-likelihood parameter estimation for unsupervised stochastic model-based image segmentation. *IEEE Trans. Image Processing*, 3(4):404–420, 1994.

[32] G. Boccignone, V. Caggiano, P. Napoletano and M. Ferraro. Image segmentation via multiresolution diffused expectation-maximisation. *IEEE ICIP'05*, 1:289–92, 2005.

[33] Joachim Weickert. Theoretical foundations of anisotropic diffusion in image processing. *Computing, Suppl.*, 11:221–236, 1996.

[34] S. B. Chaabane, M. Sayadi, F. Fnaiech and E. Brassart. Color image segmentation based on Dempster-Shafer evidence theory. *Proc. IEEE MELECON'08.*, :862–866, 2008.

[35] S. B. Chaabane, F. Fnaiech, M. Sayadi and E. Brassart. Relevance of the Dempster-Shafer evidence theory for image segmentation. *Proc. IEEE SCS'09 (3rd)*, :1–4, 2009.

[36] S. B. Chaabane, M. Sayadi, F. Fnaiech and E. Brassart. Dempster-Shafer evidence theory for image segmentation: application in cells images. *IJICT*, 5(2):126–132, 2009.

[37] Y. Neng-Hai and Y. Yong. Multiple level parallel decision fusion model with distributed sensors based on Dempster-Shafer evidence theory. *Int. Conf. Mach. Learn. Cyber.*, 5:3104–3108, 2003.

[38] S. Salicone. *Measurement Uncertainty: An Approach via the Mathematical Theoy of Evidence.* Springer, 2006.

[39] N. L. Johnson, S. Kotz and N. Balakrishnan. *Continuous Univariate Distributions*, 1. Wiley-Interscience, 2nd edition, 1994.

[40] S. Sanjay-Gopal and T.J. Hebert. Bayesian pixel classification using spatially variant finite mixtures and the generalized EM algorithm. *IEEE Trans. Image Process.*, 7(7):1014–1028, 1998.

[41] M. Gürtler, J. P. Kreiss and R. Rauh. A non-stationary approach for financial returns with nonparametric heteroscedasticity. Technical report, Institut für Finanzwirtschaft, Technische Universität Braunschweig, Sep. 2009.

[42] Y. Delignon, A. Marzouki and W. Pieczynski. Estimation of generalized mixtures and its application in image segmentation. *IEEE Trans. Image Processing*, 6(10):1364–1375, 1997.

[43] L. Liu and R. R. Yager. Classic works of the Dempster-Shafer theory of belief functions: An introduction. *Classic Works of the Dempster-Shafer Theory of Belief Functions*, :1–34, 2008.

[44] T. Denoeux. A *k*-nearest neighbor classification rule based on Dempster-Shafer theory. *IEEE Trans. Systems, Man & Cybernetics*, 25(5):804–813, 1995.

[45] P. Smets. Constructing the pignistic probability function in a context of uncertainty. 5th *Annual Conf. Uncertainty in Artifical Intelligence (UAI'89)*, :29–39, 1989.

[46] M. Meilă. Comparing clustering by the variation of information. 6th *Annual Conf. Compter Learning Thoery (COLT)*, :173–187, 2003.

[47] T. Fei and D. Kraus. An evidence theory supported Expectation-Maximization approach for sonar image segmentation. *Int. Multi-Conf. Systems, Signals & Devices'12, Communication & Signal Processing*, :1–6, 2012.

# Biographies

**Tai Fei** received the B.Eng. degree in telecommunication engineering from Shanghai Maritime University, Shanghai, China, in July 2005 and the Dipl.-Ing. (M.Sc.) and the Dr.-Ing. (Ph.D) degrees in electrical engineering and information technology from Technische Universität Darmstadt (Darmstadt University of Technology), Darmstadt, Germany, in June 2009 and November 2014, respectively. From September 2009 to December 2012, he worked as a research associate with the Institute of Water-Acoustics, Sonar-Engineering and Signal-Theory (IWSS) at the Hochschule Bremen (City University of Applied Sciences), Bremen, Germany, in collaboration with the Signal Processing Group (SPG) at Technische Universität Darmstadt (Darmstadt University of Technology), Darmstadt, Germany, where his research interest has become the detection and classification of underwater mines in sonar imagery. From August 2013 to February 2014 he worked as a research associate with Center for Marine Environmental Sciences (MARUM) at Universität Bremen (University of Bremen), Bremen, Germany, where his job in the sonar signal processing supported the exploring of underwater methane gas bubbles. Since March 2014, he is working as a development engineer at Hella KGaA Hueck & Co., Lippstadt, Germany, where he is mainly responsible for the development of reliable signal processing algorithms for automotive radar systems.

**Dieter Kraus** received the Dipl.-Ing. (B.Sc.) degree from Fachhochschule Bielefeld, Germany in 1983 and the Dipl.-Ing. (M.Sc.) and the Dr.-Ing. (Ph.D) degrees from Ruhr University Bochum, Germany, in 1987 and 1993, respectively, all in electrical engineering. In 1987 he attended the SACLANTCEN La Spezia, Italy as a Summer Research Assistant. From 1987-1992 he was research associate at the Institute of Signal Theory, Ruhr University Bochum and visiting research associate at ETSI Madrid, CHEPAG-ENSIEG Grenoble, CAPS-IST Lisbon. From 1993-2001 he was with ATLAS ELEKTRONIK, Bremen, where he was finally head of the Department for Future Mine Hunting Sonar Concepts. Since 2001, he is Professor of Signal Processing with applications in Technical Acoustics at the Hochschule Bremen, Germany. His main research interests lie in the area of Underwater Acoustics, particularly in the design of acoustic transmitter and receiver arrays, data acquisition and signal conditioning techniques, advanced sonar signal and image processing methods.

E. Markert, M. Shende, T. Horn, P. Wolf and U. Heinkel

# Tool-Supported Specification of a Wind Turbine

**Abstract:** Requirements management systems are a key technology to raise the efficiency of a design process. This paper presents the current opportunities of the tool SpecScribe in different physical domains to assist the development of a wind turbine health monitoring system. SpecScribe not only offers document organization but also direct design support by code generation and formal verification issues. This includes digital and analog problems which are transformed to FSM and hybrid automata. From these automata a SystemC-AMS model can be generated automatically. Furthermore a VHDL-AMS model is included in the specification as "golden model".

**Keywords:** Specification, Modeling, SystemC-AMS, VHDL-AMS, Wind energy generation, Refining.

# 1 Introduction

The system complexity is rising following Moore's law, forcing the design process to become more quickly to handle the emerging technology features. Design process using top-down methodology starts with the system specification. Today this is usually a collection of informal texts or pictures. It is a hard and time-consuming process to ensure the consistency and unambiguity of informal language. Several Requirement Management Systems (RMS) like DOORS or HP Quality Center help the designer teams during design process. But these tools lack of direct design support (see section 2 for details).

The tool SpecScribe will fill this gap by combining requirement management features with modeling and code generation options. Section 3 gives an overview on the currently available and planned features. The goal of the tool is to reduce the Time-to-Market of products by assisting the designer teams in recurring work items and by reducing delays caused by ambiguities from informal specifications. SpecScribe is not meant to be "yet another document management system" as it includes formal specification methods and interfaces to common modeling languages like VHDL and SystemC(-AMS) together with formal languages like PSL.

In this paper SpecScribe is used for the specification of a wind turbine system with an included structural health monitor in its rotor blades. This system as described

**E. Markert, M. Shende, T. Horn, P. Wolf and U. Heinkel:** Chair Circuit and Systems Design, Chemnitz University of Technology, Chemnitz, Germany, email: erik.markert@etit.tu-chemnitz.de.

De Gruyter Oldenbourg, ASSD – Advances in Systems, Signals and Devices, Volume 4, 2017, pp. 83–95.
DOI 10.1515/9783110448399-008

in section 4 is currently under development. The specification process of the wind turbine is shown in section 7. The formalized parts of the specification are used to generate a SystemC-AMS [1] representation of the system. The paper closes with the conclusion section and a short outlook.

# 2 Related work

Requirements Management (RM) tools have become generally established in design processes in the last years. Depending on the specific demands of a project, different aspects have to be considered. In [2], a collection of criteria for RM tools is presented.

Generally, there is a wide range of tasks that these tools are able to accomplish, and hence used for. Requirements can be tracked, i.e. every change to a requirement is recorded, including information such as who made a change and when was it made. It is also a common feature to trace dependencies between requirements and provide a visualization of these dependencies. Verification and testing are important aspects during design, as they are necessary to determine whether requirements are met. This is reflected by providing the possibility to link test cases and test plans to certain requirements, thus establishing a direct connection between specification and testing.

There are classical tools like IBM Rational Doors [3], that help designers to clearly organize requirements in a defined and well-arranged way. Other tools focus on Application Lifecycle Management (ALM) and provide functionality to cover aspects like testing, bug reporting and tracking or continuous application development. Examples for this kind of tools are HP Quality Center [4], MKS Integrity [5] or Siemens Teamcenter [6]. Unfortunately these tools not fully support a classical electronic design flow using SystemC or VHDL and have limited capabilities for formal verification. The tool SpecScribe tries to fill this gap.

# 3 The tool SpecScribe

The tool named SpecScribe, a development of our chair, supports the engineer in the specification, design and verification process. The concepts of the tool were described earlier [7], now the main parts of the tool are implemented and applied to a real world example. The data given by the specification engineers usually consist of a huge amount of non-formal textual or graphical descriptions sometimes even without a structure. Due to the textual form these descriptions often are ambiguous or inconsistent. The first step in formalization is therefore the structuring of the requirements in a hierarchical order to identify dependencies. In the next step the textual/graphical requirements need to be formalized which requires a differentiation regarding the time-continuous and digital parts.

Digital behavior means a discretisation of the time axis so a step-by-step execution of these algorithms is possible. This allows the usage of finite state machines (FSM) in ADeVA notation [8] with a state-based model of computation. Once an FSM is present in a design, model checking tools can be used to prove the system correctness. Behavioral components like FSM can be arranged hierarchically in structured components. The subcomponents communicate via signals and allow a partitioning of the system.

The analog electrical and non-electrical behavior can be formally specified defining either a reference signal form (sensor transfer function) including tolerances or a non-linear behavior using hybrid automata. Hybrid Automata [9] are an extension to digital finite state machines with time-continuous behavior taking place inside the states. Additionally information about electrical circuits can be saved. This includes the electrical connection data as well as chip packaging information for PCB design. This information will be used to perform plausability checks like level comparisons (5V vs. 1.8V) in one of the next versions of the tool. In digital domain these checks can already be performed using properties. SpecScribe currently offers interfaces to PSL and NuSMV, others are planned.

To gain acceptance of the system level designers an easy-to-use GUI is necessary. Figure 1 shows the GUI of SpecScribe. On the left hand side the specification structure is visualized where new items can be appended. The right hand side includes the detailed information of the chosen specification item. In the figure a textual requirement is shown. Requirements can be extended by subrequirements like the "blade" requirement.

The three-layered architecture of SpecScribe with structural components and requirements on the inner (first) layer, behavioral description on the second layer and tool interaction (code export/import, model checker interaction) on the outer (third) layer is realized using the platform independent languages Python and Qt. The specification data is collected using a hierachical structure of requirements which can be refined and lead to digital or hybrid automata. For continuation of the design process several code generators are available. Pure digital systems may be exported to VHDL and SystemC while analog (electrical and non-electrical) parts can be converted to a SystemC-AMS model. SystemC-AMS [10] is an extension library to SystemC (IEEE Std. 1666) for modeling analog and mixed signal behavior at system and algorithmic level and currently under standardization.

Other big issues of RMS are tracking and tracing. This means on the one hand to show who has changed what at which time. On the other hand tracing means connecting requirements and implementations to keep in mind why the implementation was made exactly this way. The tracing of changes can be done automatically using the database interface. The latter one is supported directly by SpecScribe. Several types of dependencies can be chosen like *implements* or *verifies*. To observe the status of the design project all the specification items include a satisfaction state. This state describes the current fulfillment of the requirements in different granularities like boolean or as a percentage value.

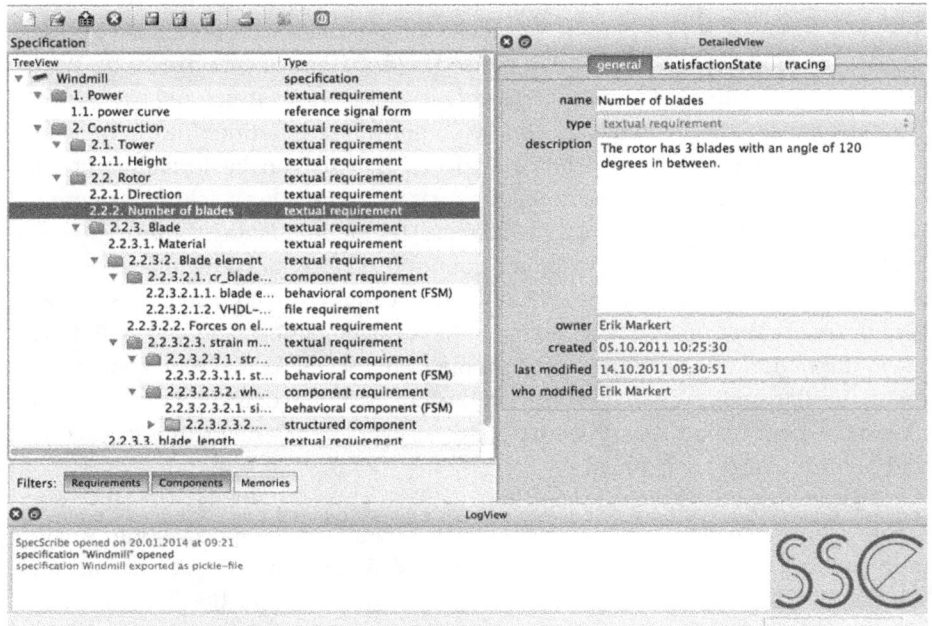

**Fig. 1.** Screenshot of wind turbine specification in SpecScribe tool.

# 4 Wind turbine system

Wind turbines are subjected to very specific loads and stresses. As the nature of the wind is highly variable, the loads acting on the wind turbine also vary greatly. Varying loads are far more difficult to analyze than the static loads because of the material aging and fatigue. Large wind turbines are inevitably elastic and the varying loads thus create a complex aeroelastic interplay which induces vibrations and resonances and can produce high dynamic load components [11]. The sources of all the forces acting on the rotor can be categorized as follows:

- aerodynamic loads
- gravitational loads
- inertial loads (including centrifugal and gyroscopic loads), and
- operational loads arising from actions of the control system, for example loads generated during braking, yawing, blade pitch control, electrical loading/unloading, etc.

Under this project, a real time Structural Health Monitoring System (SHMS) is being developed. It mainly consists of a set of electrical strain gauge sensors, micro-machined acoustic emission sensors and an intelligent system which is capable of analyzing the real time data and generate appropriate control signals. Since the strain gauge sensor modeled in this paper is developed to sense the stresses, which get generated in the plane of rotation of the rotor disc, only these forces are discussed in the next subsections.

## 4.1 Aerodynamic loads

Blade Element Momentum (BEM) theory is the most frequently used mathematical model to design a rotor blade and to evaluate various loads acting on the rotor disc. This allows the rotor blade to be analyzed in sections, where the individual forces over different sections are summed up to get the resultant force acting on the rotor. Figure 2 shows the cross section of the modeled blade. The blade is 2.1 meter long and for BEM analysis it has been divided into 5 sections. Moment force $F_M$, Radial force $F_R$ and gravitational force $F_G$ acting on third element are shown. The direction of $F_M$ depicts the direction of rotation of the rotor, here anti-clockwise. With $r$ being radial distance of the element from rotor hub and $n$ being rated revolutions per second (rps) of the rotor, the linear velocity $u$ associated with a blade element is given by:

$$u = 2\pi r n$$

In modern wind turbines, the direction of the wind $v$ and direction of the linear velocity $u$ are perpendicular to each other. The airfoil experiences the wind flow in the direction of $w$, which is known as relative wind. Figure 3 shows the vector diagram representation of various aerodynamic components acting on an airfoil. The lift force $F_L$ and the drag force $F_D$ act parallel and orthogonal to $w$ respectively and are given by the following equations [12]:

$$F_L = \frac{1}{2} \rho A_b w^2 C_L \tag{1}$$

$$F_D = \frac{1}{2} \rho A_b w^2 C_D \tag{2}$$

where $\rho$ is the air density, $A_b$ is the area of blade element. $C_L$ and $C_D$ are the life and drag coefficients of the airfoil profile and are given by the blade manufacturers in the airfoil data-sheet.

In Fig. 3, it can be seen that the resultant force $F$ can be resolved in horizontal component $F_T$ and vertical component $F_M$. $F_T$ gives the value of the thrust force and $F_M$ gives the value of the moment producing force or the torque force acting on the rotor blade. These forces are given as per the following equations:

$$F_T = F_L \cos I + F_D \sin I \tag{3}$$

$$F_M = F_L \sin I - F_D \cos I \tag{4}$$

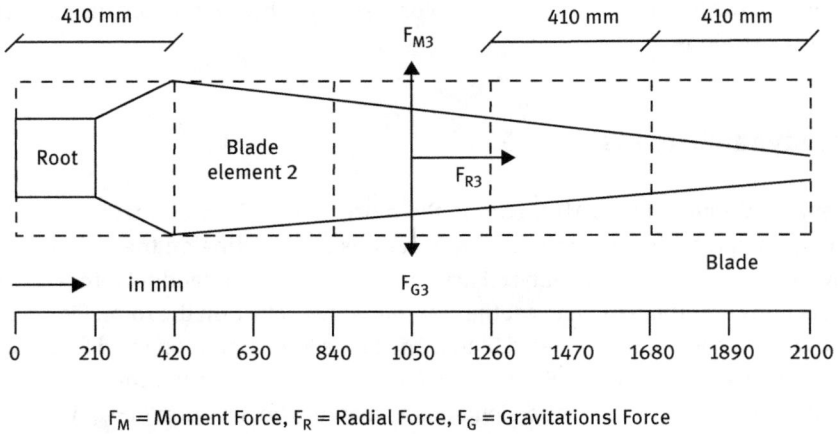

$F_M$ = Moment Force, $F_R$ = Radial Force, $F_G$ = Gravitationsl Force

**Fig. 2.** Various forces acting in the plane of rotation on the rotor blade.

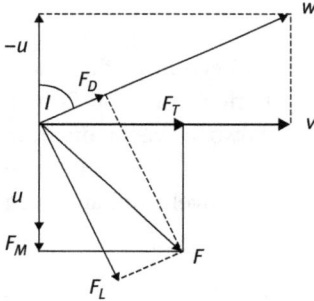

**Fig. 3.** Velocities and forces associated with an airfoil.

## 4.2 Gravitational loads

Gravity loading on the blade results in a sinusoidally varying edgewise bending moment which reaches a maximum when the blade is horizontal, and which changes its sign from one horizontal position to the other. It is thus a major source of fatigue loading [13]. Total gravitational loading is equal to the total weight of the blade.

## 4.3 Centrifugal loads

For a rigid blade rotating with its axis perpendicular to the axis of rotation, the centrifugal force also known as radial force generates a simple tensile load in the blade. In the case of small size wind turbines (diameter < 10 m), centrifugal force largely contributes to the net forces acting on the rotor blade as compared to the medium and large size wind turbines due to their higher rated speeds. For a blade element with mass $M$, the radial force $F_R$ is given by [14]:

$$F_R = M \, r \, (2\pi n)^2 \tag{5}$$

# 5 Principle of operation of strain gauge

If a strip of electrical conductor is stretched or compressed, its diameter and area of cross section will change. This geometrical variation changes its end-to-end electrical resistance. Electrical wire strain gauge uses this property of the electrical conductor and senses the strains produced in the material.

Consider a wire strain gauge. The resistance of a wire $R$ with resistivity $\rho$, length $L$ and cross-section area $A$ is given by:

$$R = \rho \frac{L}{A} \tag{6}$$

Therefore the change in resistance $dR$ is the combined effect of change in length, cross-section area and resistivity.

$$\frac{dR}{R} = \frac{dL}{L} - \frac{dA}{A} + \frac{d\rho}{\rho} \tag{7}$$

In a strain gauge, the change in electrical resistance is a function of the applied strain $\varepsilon$ and the sensitivity factor $S$ and is given by [15]:

$$\frac{dR}{R} = S \, \varepsilon \tag{8}$$

when a strain gauge is bonded well to the surface of an object, it also deforms with the deformation in the object. For a given material, the sensitivity of the material versus strain can be calibrated by the following equation [15].

$$S = 1 + 2\nu + \frac{d\rho/\rho}{\varepsilon_L} \tag{9}$$

where $\nu$ is Poisson's ration of the conductor material.

If the sensitivity factor is known, the average stain in the strain gauge can be obtained by measuring the change in electrical resistance of the gauge.

$$\varepsilon_L = \frac{dR/R}{S} \approx \frac{\Delta R}{SR} \tag{10}$$

# 6 VHDL-AMS Model of strain gauge sensor

The VHDL-AMS model presented in this section is a classical example of top-down modeling approach which exploits behavioral and structural modeling features of the language. As shown in Fig. 4, the blade element model is divided into four main components: 1. blade element model, 2. stiffness model, 3. strain gauge sensor model and 4. Wheatstone bridge model. All four models are designed separately using behavioral modeling approach and then the connections between them are defined using structural modeling approach. Considering the scope of this paper, only blade aerodynamic model and strain gauge sensor model have been discussed in the paper in greater details.

The sensor model described in VHDL-AMS is based on mechanical and electrical domains and uses the equations discussed in section 4 and 5. It can be seen in Fig. 4 that blade element model has one terminal in common with stiffness model and geometrical model of strain gauge sensor.

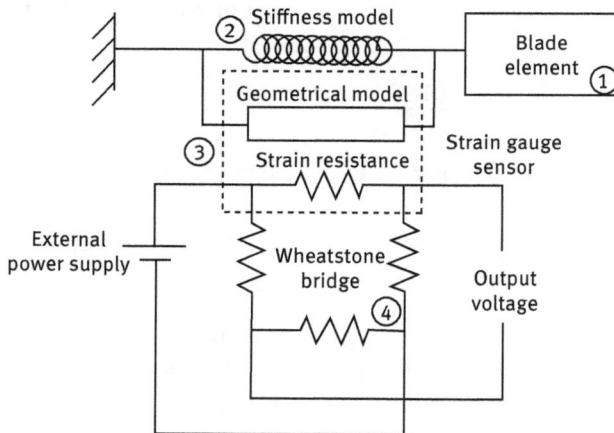

**Fig. 4.** Schematic model of a wind turbine blade element.

For modeling purpose this terminal is considered to be as translational terminal. The aerodynamic forces act on this terminal and results in a mechanical displacement. The strain gauge model has electrical as well as translational terminals. The translation terminals transfer the mechanical displacement produced by the aerodynamic model to strain gauge sensor element and electrical element produce the proportionate resistance variation, which will be further sensed through the Wheatstone bridge.

# 7 Specification process of the wind turbine

## 7.1 Creating requirements

The specification process starts with the definition of the key requirements: What should the system do and what are the basic environment conditions to be met. Environment not only includes the physical conditions of operation but also government policies, financial limitations and safety issues. These requirements need to be refined by the system engineers, leading to a collection of textual, graphical and perhaps also (from former projects) reused items. Specifications usually contain several thousands requirements, so a fast database system is necessary to provide parallel access to the data. For the wind turbine system an object oriented database is used. Larger specifications may require a relational database for faster access. This feature is currently under development.

The wind turbine example specification consists of technical requirements. If the position of install would be included in the specification also non-technical requirements caused by government rules need to be considered. Such a non-technical requirement could be: "Should not disturb the neighbors". SpecScribe allows the entry of textual requirements and the attachment of files in a flat as well as in a structured way. So a first brief (and informal) description of the system can be introduced. Usually these first requirements lack of clearness and testability. One of the main sources of errors during the specification process are ambiguous system descriptions so it is important to qualify these requirements. One solution is to pass a time-consuming review process, another solution is the formalization of the specification.

The formalization of requirements, e.g. based on the equations stated in section 4, also enables further design automation. In the wind turbine example two ways of formalization for analog behavior can be used. On the one hand the definition of Reference Signal Forms like for the system transfer function (wind velocity vs. generated power) is possible, on the other hand more complex issues can be formalized using digital or hybrid automata. To use such automatas it is necessary to map the system equations on certain states. In the wind turbine example an init state is used to assign the starting values to the variables and to introduce constants. The system behavior is formalized using special "calculation" states. The separation points for splitting up the calculations into different states are determined by complexity issues and by physical domains. Some suitable breakpoints are given in Fig. 4 and match the VHDL-AMS model structure.

## 7.2 Creating models

The key point of formalization is machine readability which allows the specification to be checked automatically in future e.g. for consistency or completeness. As addition to formalized descriptions also source code files of different hardware description languages and simulators can be included in the tool. As automatons (hybrid as well as digital) have a fixed structure it is possible to generate simulation models from the descriptions. In the wind turbine example the hybrid automaton is converted into a SystemC-AMS model. This allows a simulation and so a first verification of the system idea. Also structural models can be described in SpecScribe and generated as SystemC-AMS source code allowing complex system models as well as test benches. SystemC-AMS models can be further refined to implement the system as described in [16].

The automated model generation eliminates an important source of errors in the design process: the transfer of the specification into the first step of the implementation phase. As also the make files and a test bench framework are generated the system level designer can easily start simulation without deeper knowledge of the description language.

The included Reference Signal Forms (RSF) act as verification items. Each RSF contains a function and a tolerance range around it. So each RSF can be translated to a monitor which checks the proper system behavior by analyzing the output function. This is another step for a (semi-) automation of analog verification.

## 7.3 Results

The rotor blade also includes the health monitoring sensors as introduced in the last section. The VHDL-AMS model of the equations given there is included as "golden model" in the specification process. Figure 5 shows the part of the GUI containing the model. Such a "file requirement" can also be used to include larger texts or graphics. The GUI provides an automated link to the associated editor or viewer.

The behavior of the sensor is also formalized using a very simple hybrid automaton consisting of some of the equations given in section 4. The modeled equations are shown in Fig. 6 in the stateview of the GUI.

This automaton can be used to generate a first model in SystemC-AMS at system level. SystemC-AMS was preferred against Matlab/Simulink as an open-source implementation is available and linear electrical behavior can be included for further model refinement (e.g. to model the Wheatstone bridge). Listing of Fig. 6 shows a part of the generated SystemC-AMS model for a blade element.

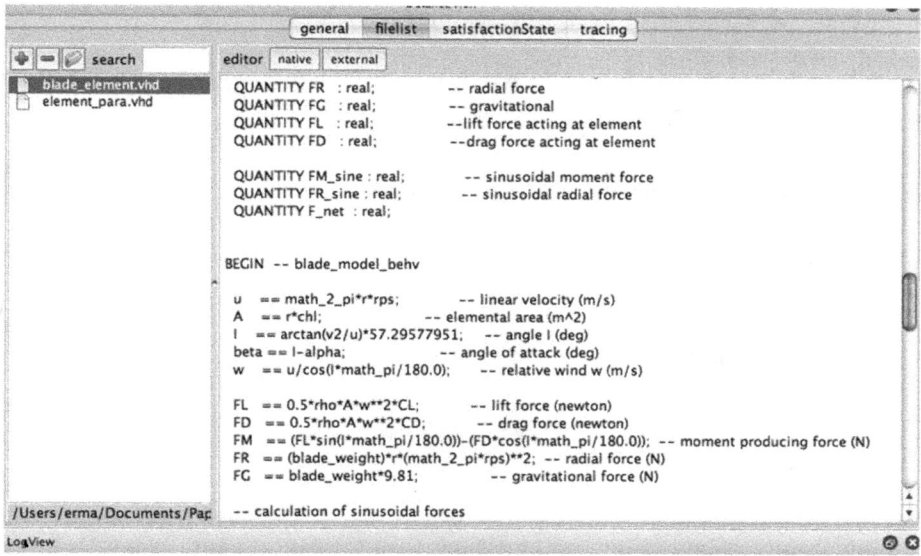

**Fig. 5.** Details of a file requirement including the VHDL-AMS model.

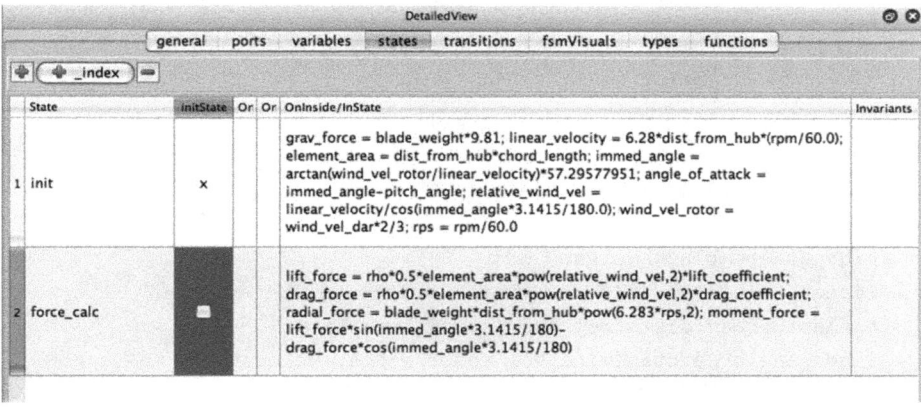

**Fig. 6.** System behavior description using SpecScribe.

```
#include "systemc-ams"
using namespace sc_core;

SCA_TDF_MODULE(blade_element_model) {
 //ports
 sca_tdf::sca_out<double > complete_force;

 //types/enums
 enum blade_element_modelstatetype { init, force_calc };

 //signals and member vars
 float blade_weight;
 float rpm;
 [...]

 void processing();

SCA_CTOR(blade_element_model) : complete_force("
    complete_force"){
 blade_weight = 0.538;
   [...]
  }
};
void blade_element_model::processing(){

 switch (blade_element_modelstate){
  case(init):
    grav_force= blade_weight * 9.81;
    linear_velocity= 6.28 * dist_from_hub * rpm / 60.0;
    element_area= dist_from_hub * chord_length;
    immed_angle= arctan(wind_vel_rotor/linear_velocity)
        *57.29577951;
    angle_of_attack= immed_angle - pitch_angle;
    relative_wind_vel= linear_velocity/cos(immed_angle
        *3.1415/180.0);
    wind_vel_rotor= wind_vel_dar*2/3;
break;
case(force_calc):
    lift_force= rho * 0.5 * element_area * pow(
        relative_wind_vel,2) * lift_coefficient;
```

```
drag_force= rho * 0.5 * element_area * pow(
    relative_wind_vel,2) * drag_coefficient;
break;
}
}
```

Simulation of these models leads to first results in system behavior verification.

By introducing SpecScribe in the design flow of the system three main advantages are noticable:

- All changes in the specification are now tracked directly in the tool.
- The hierarchical organization allows a better documentation. This documentation can be automatically written to OpenOffice data format.
- Generation of VHDL and SystemC-AMS models allows an executable specification early in the design process

# 8 Conclusion and outlook

This paper showed the specification process of a wind turbine system with integrated structural health monitoring supported by the specification tool SpecScribe. The collection of informal requirements is extended by formalized items. These items are used for automated code generation to SystemC-AMS replacing an error-prone manual conversion into an AMS simulation language.

Additionally these formal items are the base point for the formal verification of the system. Future work will focus on the verification of hybrid automata making analog and non-electrical requirements checkable at system level. This would also allow a continuous verification at different stages during the design process. The specification tool is also in use for other domains like telecommunication engineering and electromobility.

**Acknowledgment:** The work presented in this paper was done within the projects FiZ-E and SMINT. The project "FiZ-E" was funded by BMBF under the program "Forschung für den Markt im Team" (For-MaT). The project SMINT is funded by the German Research Association (DFG) within the Research Unit 1713.

# Bibliography

[1]  K. Einwich, C. Grimm, M. Barnasconi and A. Vachoux. Introduction to the systemc ams draft standard. *IEEE Int. SOC Conf.*, :446, September 2009.

[2]  M. Hoffmann, N. Kuhn, M. Weber and M. Bittner. Requirements for requirements management tools. 12th *IEEE Int. Requirements Engineering Conf.*, :301–308, September 2004.

[3]   *Getting Started with Rational DOORS*. White Paper, IBM, 2010.
[4]   Hewlett-Packard Development Company, L.P.. *HP Quality Center Enterprise Software*, data sheet, 2013.
[5]   *An Innovative Approach to Managing Software Requirements*. White Paper, MKS, 2009.
[6]   Siemens Product Lifecycle Management Software Inc. *Teamcenter: Smarter decisions, better products through end-to-end PLM*. Siemens PLM technical brochure, 2012.
[7]   U. Proß, E. Markert, J. Langer, A. Richter, C. Drechsler and U. Heinkel. A Platform for Requirement Based Formal Specification (short paper). *Forum on Specification and Design Languages (FDL)*, :237–238, September 2008.
[8]   W. Haas, U. Heinkel and S. Gossens. Semantics of a Formal Specification Language for Advanced Design and Verification of ASICs (ADeVA). *11. E.I.S.-Workshop*, April 2003.
[9]   T. A. Henzinger. The Theory of Hybrid Automata. 11th *IEEE Annual Symp. on Logic in Computer Science*, :278–292, 1996.
[10]  K. Einwich, C. Grimm, M. Barnasconi and A. Vachoux, *Standard SystemC AMS extensions Language Reference Manual*. OSCI/Accelera Std., 2010.
[11]  E. Hau. *Wind Turbines - Fundamentals, Technologies, Applications, Economics*. Springer, 2005.
[12]  G. Ingram. *Wind turbine blade analysis using the blade element momentum method*. Durham University, Tech. Rep., 2005.
[13]  T. Burton, D. Sharpe, N. Jenkins and E. Bossanyi. *Wind Energy - Hand Book*. John Wiley & Sons Ltd, 2001.
[14]  G. Sedlacik. Beitrag zum Einsatz von unidirektional naturfaserverstaerkten thermoplastischen Kunststoffen als Werkstoff fuer grossflaechige Strukturbauteile. Ph.D. dissertation, Chemnitz University of Technology, 2003.
[15]  R. L. Hannah and S. E. Reed. *Strain gauge users' handbook*. Elsevier Science Publishers Ltd, 1994.
[16]  E. Markert, M. Dienel, G. Herrmann and U. Heinkel. SystemC-AMS assisted design of an Inertial Navigation System. *IEEE Sensors J., Special Issue on Intelligent Sensors*, 7(5):770–777, May 2007.

# Biographies

**Erik Markert** received his diploma in 2004 in Information and Communication Technology at Chemnitz University of Technology. In 2010 he completed his PhD "High Level Development of Microsystems" at Chemnitz University of Technology. His main research areas are: System level design and verification of micro and nano systems, simulation with VHDL-AMS and SystemC-AMS.

**Milind Shende** has completed his Bachelor of Electrical Engineering from University of Pune (India) in 2005 and Master of Science in Microsystems Engineering from Hochschule Furtwangen (Germany) in 2008. He is working as a member of scientific staff (Wissenschaftlicher Mitarbeiter) at the Chair 'Circuit and System Design' at 'Chemnitz University of Technology' since 2009. His areas of interest are modeling of heterogenous systems, implementation of co-simulation environment for mixed signal systems using interconnection and synchronization of multiple modeling tools.

**Thomas Horn** received his diploma in Information and Communication Technology in 2008. Since this time he works as research engineer with focus on formal methods at Chemnitz University of Technology.

**Peter Wolf** passed his diploma in electrical engineering in 2006 at Chemnitz University of Technology. He finished his PhD in 2012 with the title "Measurement system for supervision of Fibre-reinforced plastics". He currently works as a research engineer with focus on sensing and supervising technologies.

**Ulrich Heinkel** is head of the chair Circuit and System Design at Chemnitz University of Technology. He received his diploma and PhD in electrical engineering from Friedrich–Alexander–University of Erlangen–Nuremberg. Since 1999 he worked as research engineer at Lucent Technologies Nuremberg. His research focusses on formal methods, specification and verification of digital systems.

www.ingramcontent.com/pod-product-compliance
Lightning Source LLC
Chambersburg PA
CBHW081109220326
41598CB00038B/7285